1 MONTH OF
FREE
READING

at

www.ForgottenBooks.com

By purchasing this book you are
eligible for one month membership to
ForgottenBooks.com, giving you
unlimited access to our entire
collection of over 1,000,000 titles via
our web site and mobile apps.

To claim your free month visit:

www.forgottenbooks.com/free904847

ISBN 978-0-266-88763-8
PIBN 10904847

This book is a reproduction of an important historical work. Forgotten Books uses
state-of-the-art technology to digitally reconstruct the work, preserving the original format
whilst repairing imperfections present in the aged copy. In rare cases, an imperfection in
the original, such as a blemish or missing page, may be replicated in our edition. We do,
however, repair the vast majority of imperfections successfully; any imperfections that
remain are intentionally left to preserve the state of such historical works.

NBS TECHNICAL PUBLICATIONS

PERIODICALS

JOURNAL OF RESEARCH reports National Bureau of Standards research and development in physics, mathematics, chemistry, and engineering. Comprehensive scientific papers give complete details of the work, including laboratory data, experimental procedures, and theoretical and mathematical analyses. Illustrated with photographs, drawings, and charts.

Published in three sections, available separately:

⊙ **Physics and Chemistry**

Papers of interest primarily to scientists working in these fields. This section covers a broad range of physical and chemical research, with major emphasis on standards of physical measurement, fundamental constants, and properties of matter. Issued six times a year. Annual subscription: Domestic, $9.50; foreign, $11.75*.

⊙ **Mathematical Sciences**

Studies and compilations designed mainly for the mathematician and theoretical physicist. Topics in mathematical statistics, theory of experiment design, numerical analysis, theoretical physics and chemistry, logical design and programming of computers and computer systems. Short numerical tables. Issued quarterly. Annual subscription: Domestic, $5.00; foreign, $6.25*.

⊙ **Engineering and Instrumentation**

Reporting results of interest chiefly to the engineer and the applied scientist. This section includes many of the new developments in instrumentation resulting from the Bureau's work in physical measurement, data processing, and development of test methods. It will also cover some of the work in acoustics, applied mechanics, building research, and cryogenic engineering. Issued quarterly. Annual subscription: Domestic, $5.00; foreign, $6.25*.

TECHNICAL NEWS BULLETIN

The best single source of information concerning the Bureau's research, developmental, cooperative and publication activities, this monthly publication is designed for the industry-oriented individual whose daily work involves intimate contact with science and technology—*for engineers, chemists, physicists, research managers, product-development managers, and company executives.* Annual subscription: Domestic, $3.00; foreign, $4.00*.

* Difference in price is due to extra cost of foreign mailing.

NONPERIODICALS

Applied Mathematics Series. Mathematical tables, manuals, and studies.

Building Science Series. Research results, test methods, and performance criteria of building materials, components, systems, and structures.

Handbooks. Recommended codes of engineering and industrial practice (including safety codes) developed in cooperation with interested industries, professional organizations, and regulatory bodies.

Special Publications. Proceedings of NBS conferences, bibliographies, annual reports, wall charts, pamphlets, etc.

Monographs. Major contributions to the technical literature on various subjects related to the Bureau's scientific and technical activities.

National Standard Reference Data Series. NSRDS provides quantitive data on the physical and chemical properties of materials, compiled from the world's literature and critically evaluated.

Product Standards. Provide requirements for sizes, types, quality and methods for testing various industrial products. These standards are developed cooperatively with interested Government and industry groups and provide the basis for common understanding of product characteristics for both buyers and sellers. Their use is voluntary.

Technical Notes. This series consists of communications and reports (covering both other agency and NBS-sponsored work) of limited or transitory interest.

Federal Information Processing Standards Publications. This series is the official publication within the Federal Government for information on standards adopted and promulgated under the Public Law 89–306, and Bureau of the Budget Circular A–86 entitled, Standardization of Data Elements and Codes in Data Systems.

CLEARINGHOUSE

The Clearinghouse for Federal Scientific and Technical Information, operated by NBS, supplies unclassified information related to Government-generated science and technology in defense, space, atomic energy, and other national programs. For further information on Clearinghouse services, write:

Clearinghouse
U.S. Department of Commerce
Springfield, Virginia 22151

Order NBS publications from: Superintendent of Documents
Government Printing Office
Washington, D.C. 20402

UNITED STATES DEPARTMENT OF COMMERCE
Maurice H. Stans, Secretary
NATIONAL BUREAU OF STANDARDS ● Lewis M. Branscomb, Director

TECHNICAL NOTE 392

ISSUED OCTOBER 1970

Nat. Bur. Stand. (U.S.), Tech. Note 392, 182 pages (Oct. 1970)
CODEN: NBTNA

The Thermodynamic Properties of Compressed Gaseous and Liquid Fluorine*

Rolf Prydz and Gerald C. Straty

Cryogenics Division
Institute for Basic Standards
National Bureau of Standards
Boulder, Colorado 80302

*This work was carried out at the National Bureau of Standards under the sponsorship of the United States Air Force (MIPR No. FO 4611-70-X-0001).

NBS Technical Notes are designed to supplement the Bureau's regular publications program. They provide a means for making available scientific data that are of transient or limited interest. Technical Notes may be listed or referred to in the open literature.

TABLE OF CONTENTS

TABLE OF CONTENTS -- Continued

TABLE OF CONTENTS -- Continued

TABLE OF CONTENTS--Continued

LIST OF TABLES

LIST OF FIGURES -- Continued

LIST OF SYMBOLS AND UNITS

A_i, B_i, C_i	Coefficients of various equations.
$B(T)$	Second virial coefficient, $1/mol$
$C(T)$	Third virial coefficient, $(1/mol)^2$
$C_p(T, \rho)$	Specific heat at constant pressure, $J/mol\ K$
$C_p^o(T)$	Ideal gas specific heat at constant pressure, $J/mol\ K$
$C_{sat}(T)$	Specific heat of saturated liquid, $J/mol\ K$
$C_v(T, \rho)$	Specific heat at constant volume, $J/mol\ K$
$C_v^o(T)$	Ideal gas specific heat at constant volume, $J/mol\ K$
E	Expansion coefficient
E_T	Thermal expansion coefficient of dead-weight gage piston-cylinder assembly
g	Gravity at NBS Boulder Laboratory = $9.79615\ N/kg$
g_s	Standard gravity = $9.80665\ N/kg$
$H(T, \rho)$	Enthalpy, J/mol
ΔH	Heat of vaporization, J/mol
K	Thermal conductivity, $W/m\ K$
L_c	Cryostat capillary length = $0.71\ m$
L_m	Height of mercury manometer column, m
L_n	Height of nitrogen column in intermediate pressure system = $1.04\ m$

LIST OF SYMBOLS AND UNITS -- Continued

M	Molecular weight of fluorine = 37.9968 g/mol
N	Number of g moles of fluorine
N_i	Burnett apparatus cell constant
P	Pressure, MN/m^2 (1 MN/m^2 = 10^6 newton/m^2 = 9.86923 atm)
P_b	Barometric pressure, MN/m^2
P_c	Critical pressure = 5.215 MN/m^2
P_n	Nominal dead-weight gage pressure, MN/m^2
P_t	Triple point pressure = 2.52 x 10^{-4} MN/m^2
R	Universal gas constant = 8.3143 J/mol K
$S(T, \rho)$	Entropy, J/mol K
T	Absolute temperature, degrees Kelvin, on the International Practical Temperature Scale of 1968
T_c	Critical point temperature = 144.31 K
T_b	Normal boiling point temperature = 84.950 K
T_t	Triple point temperature = 53.481 K
T_o	Pipet sample holder temperature, K
t_m	Mercury manometer temperature, °C
$U(T, \rho)$	Internal energy, J/mol
V	Molar volume, l/mol
V_{dg}	Volume of diaphragm cell and capillary at same temperature during gasometer measurements, cm^3

LIST OF SYMBOLS AND UNITS -- Continued

V_n	Volume of metal necks of gasometer flasks, cm^3
V_p^o	Normal pipet volume at zero pressure, cm^3
V_o	Normal pipet volume at zero pressure and 20°C, cm^3
W	Weight factor
W_s	Velocity of sound, m/s
Z	Compressibility factor $= PV/RT$
Z_o	Compressibility factor of fluorine in pipet at P_g and T_o
α, γ, δ, σ, χ, u, X	Statistical mechanical parameters; for definitions see References [52, 53]
θ	Cryostat capillary factor
λ	Volume ratio from Equation (7)
ρ	Density, mol/l
ρ_c	Critical density $= 15.10$ mol/l
ρ_m	Mercury density from Equation (15)
$\rho_{melt\,L}(T$ or $P)$	Density of liquid along liquid-solid boundary, mol/l
ρ_n	Nitrogen density in intermediate pressure system from Equation (17)
$\rho_{sat\,L}(T)$	Saturated liquid density, mol/l
ρ_t	Liquid triple point density $= 44.86$ mol/l

Subscripts:

	Conditions in cryostat capillary
cr	Conditions in capillary from cryostat to gasometer
d	Fluorine diaphragm cell conditions
g	Conditions in gasometer flasks
m	Gasometer manifold conditions
p	Pipet sample holder property
q	Quartz bourdon tube conditions
r	Instrument room conditions
w	Dead-weight gage conditions

calc	Calculated value
exp	Experimental value
g	Gas property
ℓ	Liquid property
o	Reference state property
sat	Properties at vapor-liquid saturation boundary

Superscript:

| | Ideal gas property |

THE THERMODYNAMIC PROPERTIES OF COMPRESSED·
GASEOUS AND LIQUID FLUORINE

Rolf Prydz and Gerald C. Straty

ABSTRACT

An apparatus has been constructed and used successfully to measure vapor pressure and PVT data of fluorine from the triple point to 300 K at pressures to about 24 MN/m². Material problems caused by the toxic and corrosive nature of fluorine were solved. A network of isotherm and isochore polynomials and a truncated virial equation were used to represent all PVT data. These equations represent the data with an average standard deviation of about 0.02 percent in density, the corresponding accuracy being estimated at 0.1 percent. Equations for the saturated liquid and vapor densities, the vapor pressure curve, the melting line, and the ideal gas properties are also presented. Comparisons are given to published values of the second virial coefficients, vapor pressures, and saturation densities. Additional comparisons are also made to measured specific heats and latent heats of vaporization. New values are reported for the triple point and critical point parameters together with the temperature and saturation densities at the normal boiling point. Finally, extensive tables of thermodynamic properties of fluorine are given which include pressure, temperature, density, isotherm and isochore derivatives, internal energy, enthalpy, entropy, specific heats at constant pressure and volume and velocity of sound.

Key Words: Density; enthalpy; entropy; fixed points (PVT); fluorine; Joule-Thomson; latent heat; melting curve; PVT measurements; saturation densities; specific heats; vapor pressure: velocity of sound; virial coefficients.

* This work was carried out at the National Bureau of Standards under the sponsorship of the United States Air Force (MIPR No. FO 4611-70-X-0001).

I. INTRODUCTION

Current increased technical and scientific interest in fluorine arises from the fact that this cryogenic fluid is being considered as a propellant oxidizer in rocket propulsion systems. The reason is that liquid fluorine is one of the most energetic oxidizers known and its use with hydrogen as a fuel results in the most favorable specific impulse of any stable oxidizer-fuel combination. Since the product of the reaction is hydrofluoric acid, the use of this combination is limited to upper stages of the rocket.

The highly toxic and reactive properties of fluorine have, however, discouraged research in determining its physical properties which are required for engineering calculations and systems development. For example, a few parts per million of fluorine in the air is fatal to humans. Also, hydrofluoric acid resulting from fluorine reactions in poorly cleaned systems may actually react with any metal used as the containment for this reactive fluid. As a result, physical property data on fluorine tend to be quite inaccurate, extremely scarce, and in many cases, nonexistent.

The primary objective of this research program was, therefore, to determine the complete PVT relationship for gaseous and liquid fluorine to be used for calculating thermal functions. These PVT measurements are also required for future experimental fluorine measurements in this laboratory of thermodynamic and transport properties such as specific heats (which have been completed at the time this is being written), velocity of sound, dielectric constant, viscosity, and thermal conductivity. Valuable knowledge concerning intermolecular forces may be gained from the PVT measurements; for instance, information about the inter-molecular potential can be obtained from the temperature depend-ence of the second virial coefficient. Further, correlation of these PVT data to calculate thermodynamic properties also re-quires vapor pressure data from the triple point to the critical point presented in the form of a vapor pressure equation. Finally, to make technical design of fluorine systems possible, extensive tables of thermodynamic properties of the saturated and single-phase regions are essential.

Although reports on some specific phases of this research have been published, it is the purpose of this note to present, under a single cover, the complete and detailed results of the PVT experimental and calculational program. Extensive tables of PVT prop-erties and derived thermodynamic functions are given. The experi-mental method and apparatus are discussed and calculational proce-dures for data correlations and thermodynamic properties computa-tions are presented.

II. EXPERIMENTAL PVT METHODS

Several different methods for accurate PVT data measurements are used in various laboratories throughout the world. While the procedures for measuring pressure and temperature are standard and vary little from laboratory to laboratory, drastically different methods are employed in obtaining the corresponding specific volume or density. A thorough review of these PVT measurement methods is given by Sengers [1].*

In general there are four different procedures that are used to obtain the specific volume, namely the Piezometer Method, the Weight Method, the Isochore Method, and the Gas Expansion Method. The latter method, which is the most widely used one, will be discussed in greater detail than the other methods since it is the approach used for this work.

A.   The Piezometer Method.

This procedure uses a piezometer of known volume and a calibrated stem as shown in Figure 1.a. It is filled with the fluid under study to a pressure of 1 atmosphere and controlled at a temperature where the isotherms of the fluid are known. Thus the total amount of the fluid (the normal volume) in the piezometer is available. A capillary connects the stem to a pipet of known volume. Hence, measurements of P-V points at different pipet

---

*Numbers in brackets refer to citations in the bibliography.

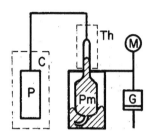

C = Cryostat
P = Pipet
Th = Thermostat
Pm = Piezometer
M = Manometer
G = Pressure Generator

Figure l.a.   The piezom-
             eter method.

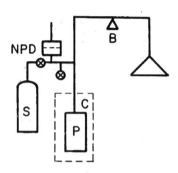

NPD = Null Pressure Detector
S = Sample Supply
C = Cryostat
P = Pipet
B = Balance

Figure l.b.   The weight method.

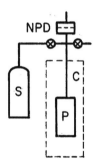

NPD = Null Pressure Detector
S = Sample Supply
C = Cryostat
P = Pipet

Figure l.c.   The isochore
             method.

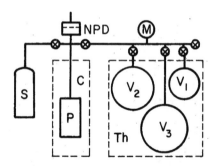

V_1, V_2, V_3 = Expansion Volumes
NPD = Null Pressure Detector
Th = Thermostat
S = Sample
C = Cryostat
P = Pipet
M = Manometer

Figure l.d.   The gas expansion
             method.

of the fluid in the pipet is easily computed from the known total sample amount and the quantity remaining in the piezometer stem which may be calculated from the known fluid properties at that temperature.

The advantage of this method is that only one normal volume determination is needed for a large range of pressures and temperatures. However, accurate volume determinations at high pressures are experimentally difficult since the ratio between volumes occupied by the fluid at 1 atmosphere and at high pressures becomes very large at low pipet temperatures. Another disadvantage of this method, which made it impractical for this work is that known P-V properties of the fluid are required at the piezometer temperature. No such data were available for fluorine.

B.   The Weight Method.

A pipet of known internal volume is suspended from a balance as shown in Figure 1.b. It is filled with a sample of fluid to a selected pressure and temperature and the increase in weight is recorded. These conditions provide the necessary information for one PVT data point.

An obvious advantage of this method is that no normal volume

this advantage. The difficulty of the sample weight determination is partly due to the fact that a compromise must be made between the freedom of motion of the pipet and its thermal contact with the surroundings. The main factor destroying the accuracy of this measurement is, however, the weight of sample being only a small fraction of the weight of the pipet.

C. The Isochore Method.

Figure 1.c is a schematic of the apparatus used in this method. A pipet of unknown volume is filled with a sample of the fluid and the pressure is measured at a temperature where the P-V relationship of the fluid is known. This permits evaluation of the specific volume of the fluid at these conditions. P-T data for that specific isochore and for other isochores may then be obtained in the same manner. Since the pressures are usually measured at the same temperatures for each isochore the resulting PVT data are rearranged in isotherms without difficulty.

The fact that no volume calibration or measurement of normal volumes are needed, makes this method quite attractive for obtaining PVT data. However, the requirement of existing accurate high-pressure data eliminated this procedure for application to fluorine since no such data were available.

D. The Gas Expansion Method.

The gas expansion method is generally used in most laboratories to obtain accurate PVT data, especially where large amounts of data are required. The time consuming calibrations of the pipet volume, the volumes of the vessels, and other noxious volumes are over-shadowed by the fact that only one normal volume determination is needed for every P-T relationship (iso-chore) measured. Some low-density isochores for fluorine re-sulted in as many as 40 PVT data points.

With reference to Figure 1.d the pipet of calibrated (calibration procedure is discussed in Chapter IV) volume in the cryostat is filled with the fluid to be studied at approximately the desired density. Then P-T points are measured at conditions of nearly constant density. When either the desired maximum temperature or pressure is attained in the pipet, the fluid is released into expansion vessels of calibrated volumes, combined in such a manner as to give a final pressure close to 1 atmosphere. This pressure is then generally measured with the aid of a manometer; in the case of fluorine a quartz bourdon tube gage was used due to the extreme reactivity of fluorine gas with manometer fluids even at low pressures. The vessel temperature is precisely con-trolled by a bath at a temperature close to the room conditions. Thus, the total amount of sample in the pipet, the cryostat capil-lary, and the null pressure detector is determined applying the gas law to the state of the sample fluid in the expansion vessels. The compressibility factor, $Z = PV/RT$, of the gas under these

as discussed in Chapter VII. To obtain the actual amount of fluid in the pipet alone, the portion of sample in the cryostat and the null pressure detector, which is dependent upon temperature, is subtracted from the total amount of sample. The experimental density at each P-T point is obtained by dividing the amount of sample in the pipet by its calibrated volume corrected for temperature and pressure effects as discussed in Chapter IV. Since the density decreases slightly with increasing temperatures along a P-T relationship, the experimental run is usually denoted as a pseudo-isochore.

This expansion method was determined to be the most practical one for fluorine for the following reasons:

1) large amounts of data could be obtained with relative ease and great accuracy;

2) no initial accurate PVT data of fluorine were required, and

3) an all metal apparatus, with the exception of the low-pressure pyrex glass expansion flasks, could be utilized minimizing the hazards of data collection. The apparatus used is discussed in the next chapter.

III. EXPERIMENTAL APPARATUS

A. Introductory Remarks.

In the early stages of fluorine technology development, a number of failures occurred that were characterized by the chemical ignition between the fluorine and its containment. The ignition zone was usually consumed in the reaction and the reason or failure was difficult to determine. It was believed, however, that either organic contamination or incompatibility of the materials was the cause. Later tests confirmed that most common metals are compatible for use in a fluorine environment, provided that the system is maintained completely free of contamination. Consequently, the high pressure section (up to 20 MN/m^2; MN/m^2 = 9.86923 atm) of the fluorine system in this study was constructed of stainless steel, copper, and brass with packless metal valves. In the low pressure section (atmospheric and below) polytetrafluoroethylene was used in a few places and the gasometry flasks were constructed of pyrex glass. The cleaning and passivating procedure of this equipment is described in detail Chapter VI.

The experimental apparatus, not including the electronics and instrumentation is shown schematically in Figure 2 and is similar to that used by Goodwin [3] for hydrogen and by Weber [4] for oxygen. However, several modifications were incorporated

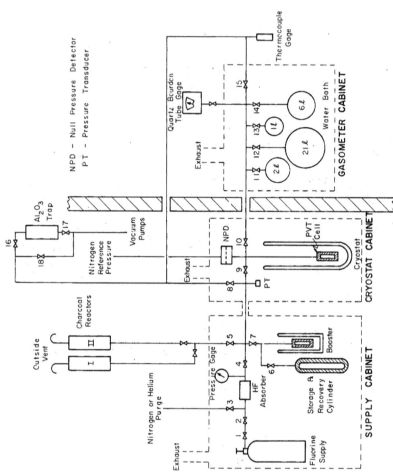

Figure 2. Fluorine PVT apparatus.

NPD – Null Pressure Detector

PT – Pressure Transducer

to minimize the problems that arise due to the toxic and reactive nature of fluorine [5]. The fluorine supply system, the cryostat, and the vacuum pumps are separated from the instrument room by an explosion-proof wall. Since fluorine pressure in the gasometer glass flasks generally is atmospheric or below, the gasometer is conveniently located in the instrument room.

All system components handling fluorine are enclosed in aluminum cabinets and vented by a separate ventilation system. Thus, in case a leak should develop, fluorine would rarely be released into the laboratories. Also, the enclosures are of sufficient strength to give reasonable blast protection from direct effects of any explosive reaction.

B. Cryostat and Pipet.

A cross-section of the metal cryostat is shown in Figure 3. The 26-cm^3 pipet sample holder, also called the PVT cell, is suspended in the lower portion of the cryostat by a thin-walled stainless steel reflux tube as pictured in Figure 4. Gaseous or liquid nitrogen or gaseous helium in this tube cools the pipet. This pipet is a 1.59 cm diameter cavity bored into a solid 5 cm diameter copper cylinder 21.6 cm long as described by Goodwin [3]. A plug was silver soldered into the lower end of the pipet sample holder while the cavity was filled with helium to avoid oxidation of the copper. Helical grooves were machined on

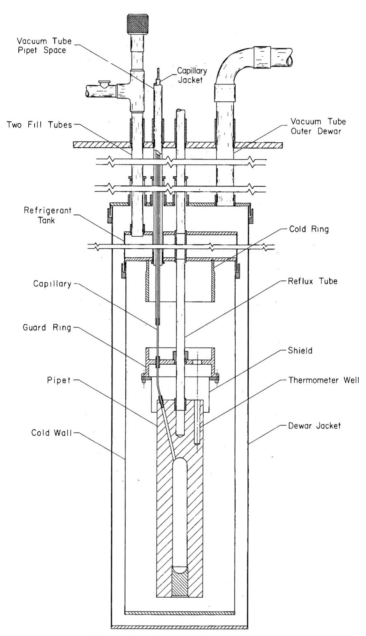

Figure 3. Cryostat and pipet.

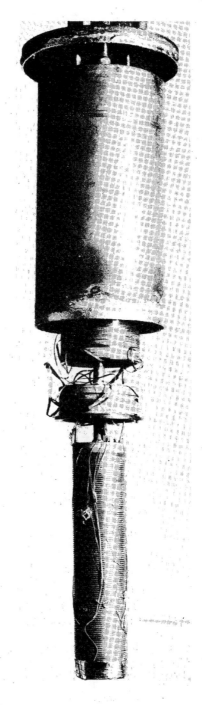

Figure 4. Refrigerant tank and pipet.

the outside of the cylinder to carry a 32-gage, 500 ohm constantan wire heater, which permits a maximum heating rate of about 1 to 2 K per minute depending on the operating temperature. The platinum resistance thermometer was tinned and secured in the thermometer well of the pipet with Rose's alloy. A guard ring is supported about 5 cm above the pipet by the reflux tube. This ring is automatically controlled to match the pipet temperature and thermally tempers the electrical leads which are wound around this copper cylinder before they are connected to the pipet. The temperature control of the guard ring is obtained by a gold-cobalt versus copper differential thermocouple between the pipet and the guard ring. A power supply provides the current for a 36-gage, 100 ohm constantan wire heater after receiving the amplified thermocouple signal. Tempering of the electrical leads entering the cryostat at room temperature is accomplished by thermally anchoring them to a cold ring suspended from the liquid refrigerant tank. Good thermal contact on the ring is insured because all wires were assembled as one unit in a parallel plane and varnished together before they were mounted.

The pipet is surrounded by an evacuable copper jacket soldered to the refrigerant tank with Rose's alloy. This jacket, which also serves as the cold wall, and the refrigerant tank are then, in turn, enclosed in another vacuum jacketed dewar. Finally, the cryostat is immersed in an open liquid nitrogen dewar having a

three-step electronic liquid level indicator. Arrangement of fore-
pump, diffusion pump and vacuum valving for the cryostat is
shown in Figure 5.

The liquid nitrogen is supplied to the inner refrigerant tank
through a permanent transfer line. A suitable valving and safety
system permits pumping on the nitrogen to attain the triple point
temperature of nitrogen and below.

Valves 9 and 10 (shown in Figure 2) on top of the cryostat
may be operated from the instrument room through appropriate
gearing and extended stems. These valves were designed by
Straty [6] for the purpose of handling corrosive fluorine at more
than 20 MN/m^2. He modified a commercially available valve,
valve A as shown in Figure 6, with bellows (valve B) such that the
primary non-metallic packing was avoided.

A 0.033 cm inside diameter stainless steel capillary connects valves 9 and 10 to the pipet. The fluorine-nitrogen separa-
tor (NPD), a commercial diaphragm differential pressure cell, is
connected between the pressure measuring system (Chapter III. E)
and the capillary at the top of the cryostat.

C.   The Gasometer.

The gasometer consists of four calibrated volumes. These
are spherical glass flasks of approximately 1 to 21 liters submerged in a precision thermostated water bath conveniently con-

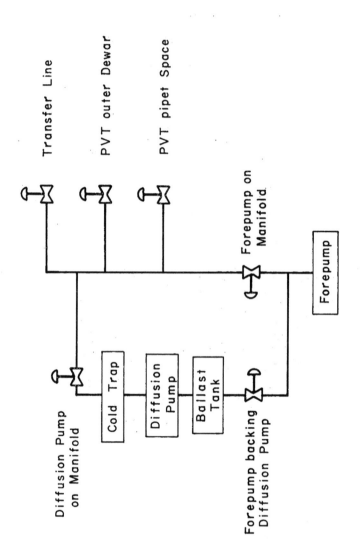

Figure 5. Cryostat vacuum valving.

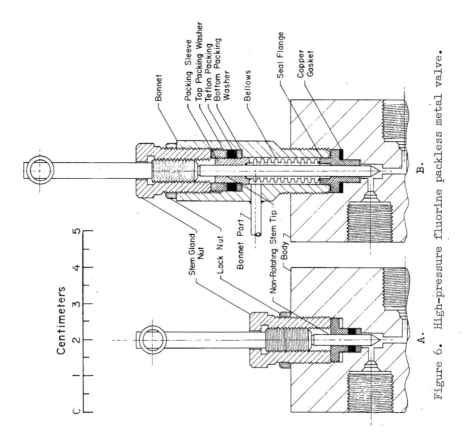

Centimeters

Figure 6. High-pressure fluorine packless metal valve.

trolled at 295.00 ± 0.01 K, the approximate instrument room tem-

perature. A higher bath temperature was not chosen since this

would have resulted in a greater bath temperature fluctuation and

excessive loss of water by evaporation into the ventilation system.

The temperature controller was adjusted according to a precision

thermometer, readable to ± 0.001 °C, and which was previously

calibrated against the pipet platinum resistance thermometer.

A 0.35 MN/m^2 fuzed quartz bourdon tube type gage with

pressure readings reproducible to 7 N/m^2, was connected to the

valve manifold. When the fluid in the pipet, cryostat capillary,

and the diaphragm cell is released into the gasometer, a standard

volume is chosen such that the final pressure in the calibrated

volumes is close to 1 atmosphere. Lower pressures will cause

a larger relative error in the pressure reading, whereas higher

pressures may break the 6- or 21-liter bulb. The quartz bourdon

tube type of gage was frequently calibrated against a high-precision

air dead-weight gage (accuracy of ± 0.015 percent of reading) to

eliminate any error that might arise due to possible changes in the

elastic properties of the quartz tube from exposure to fluorine.

Later when an apparent weakening of the tube was observed, it was

used as a null indicator and the pressure measured with the dead-

weight gage. To compromise between low noxious volumes and

adequate pumping speeds copper tubing of 0.32 cm inside diameter

old valves (valves 11, 12, 13 and 14 on Figure 2) were of the
ellow seal type and were kept fully open or closed depending on
whether the specific calibrated glass flask was to be filled or not.

1. The Fluorine Supply System.

Commercial grade fluorine gas was obtained in one half
pound cylinders pressurized to approximately 2.4 MN/m^2. This
gas was further distilled by Argonne National Laboratory to 99.99
percent purity, with the remaining impurities being mostly oxygen,
nitrogen, and hydrogen fluoride as determined from residual gas
analysis after reaction of the fluorine with mercury.

A double valving manifold system of brass needle type
valves permits fine control of the fluorine flowrate to the hydrogen
fluoride absorber (Figure 2). This HF trap consists of a section
of standard 5 cm Cu-Ni pipe equipped with caps, thermowell, and
electric heating wires. The absorber is packed with sodium bi-
fluoride ($NaHF_2$) pellets and was activated prior to installation by
heating to about 300°C while simultaneously purging with dry
nitrogen. This activation procedure drove off the hydrogen
fluoride leaving a porous absorbent form of sodium fluoride (NaF).
The trap may be reactivated in the same manner after 8 to 11 kg
of fluorine have been processed. A fluorine compatible bourdon
pressure gage indicates the pressure in the absorber at any time.

The fluorine storage and recovery cylinder is a 3-liter heavy-walled AISI 304 stainless steel cylinder having a maximum operating pressure of about 12 MN/m^2, which is five times the normal cylinder pressure. Figure 7 shows the fluorine supply and the cryostat cabinets.

Valves 4, 5, and 8 (Figure 2) are all Cu-Ni forged needle valves, whereas valves 6 and 7 are the same modified type as valves 9 and 10, described earlier and shown in Figure 6. Valves 6 and 7, however, are separate, single valves. The thermal booster is used to monitor the PVT cell filling pressure and to recover the fluorine after the gasometer measurements. The construction features of this unit were quite similar to those of the pipet. A cavity of 2.30 cm diameter was bored into a solid AISI type 304 stainless steel cylinder 4 cm in diameter. A plug was arc welded into the end of the booster while the filling line, a 0.32 cm outside diameter copper tube, was silver soldered into the top. A liquid nitrogen dewar may be raised and lowered around the cylinder through a hoist arrangement operated from the instrument room. Provisions were also available for lowering the temperature of the refrigerant bath to its triple point and below by pumping on the nitrogen dewar to recover more of the fluorine.

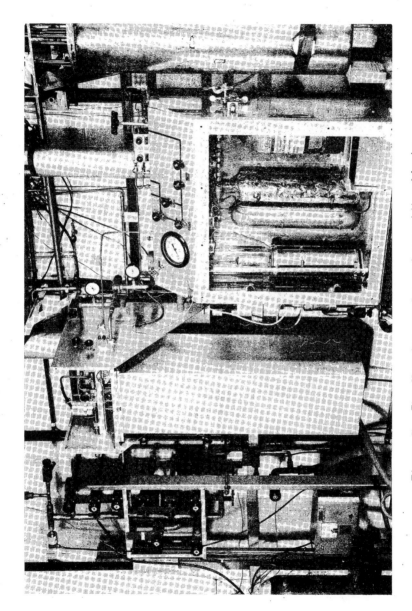

Figure 7. The fluorine supply- and cryostat cabinets.

E.   Measurement of Pressure.

1. Apparatus and Method

The commercial oil operated, dual range, precision dead-weight gage pressure measuring system, shown in Figures 8 and 9, is the same as the one used by Weber [4] and is a modification of the system used by Goodwin [3]. The system includes two null pressure detectors, often referred to as commercial diaphragm cells, with nitrogen serving as the intermediate fluid. The first null pressure detector separates the oil in the dead-weight gage from the nitrogen gas system, whereas the second separates the nitrogen from the fluorine sample. Hence, in the event of a failure of the diaphragm in the null pressure detector, high pressure fluorine gas will not come in contact with the oil to cause a hazardous and explosive condition. Even though the diaphragms are constructed of 0.005 cm thick steel sheets, they can withstand overpressures of up to 100 MN/m^2 and exhibit negligible temporary hysteresis.

The second null pressure detector shown in Figure 8 is placed on top of the cryostat to minimize noxious volumes. The rest of the pressure measuring system is located in the instrument room. The oil and the nitrogen pressures are adjusted by hand-operated screw-type pumps. With practice the two-diaphragm system may be operated with as much sensitivity as the

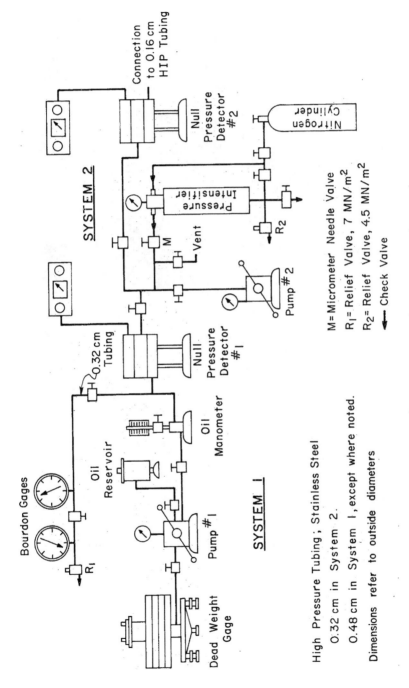

Figure 8. Schematic of the pressure measuring system.

Figure 9. Pressure measuring apparatus and gasometer.

ragm system of Reference [3].

e dual pressure ranges of the oil dead-weight gage are

by means of interchangeable piston-cylinder assemblies,

eters of which are measured to $\pm 1.3 \times 10^{-5}$ cm using a

e micrometer. The weights are equivalent to class "P"

masses. Each weight is individually adjusted to match

tive area of the piston-cylinder assembly and is calibrated

lass "S" standard masses as certified by the National

f Standards [7]. This corresponds to a 0.05 percent

on the 1 psi weight decreasing to 0.002 percent on the

weights.

e oil dead-weight gage readings are corrected for air

ʳ on the weights, the local gravity, the thermal expansion

:ic deformation of the piston and cylinder, and the baro-

ressure as discussed in detail by Cross [8].

timates of accumulated pressure errors [3] are found in

as a function of operating pressure.

TABLE 1. Estimates of Accumulated Pressure Errors

Absolute pressure, MN/m²	0.4	3	20
	Error, MN/m² x 10⁴		
Barometer	0.51	0.51	0.51
Null Pressure Detector	4.05	4.05	4.05
Piston-cylinder diameter	0.12	0.91	6.07
Calibration of weights	1.36	2.31	4.12
	6.04	7.78	14.75
Percent pressure error	0.151	0.026	0.007

2. Pressure Adjustment for Sample in Capillary

The static pressure created by the sample in the capillary column of length L_c extending from the pipet to valves 9 and 10 must be subtracted from the observed pressure to obtain the true pressure in the pipet. This static pressure head is given as

$$P_c = \frac{M L_c \theta}{R T_0} P \tag{1}$$

where

M = molecular weight of fluorine = 37.9968 g/mol

L_c = capillary column length = 71 cm

θ = capillary factor, Equation (5)

P = observed pressure, MN/m²

R = universal gas constant = 8.3143 J/mol K

T_0 = temperature of pipet, K

e the factor MH_c/R_{10} varies for

than 0.001 for more than 95 percent of the data points.

asurement of Temperature.

1. Instrumentation and Method

temperature measurement and control system is shown

10 and 11. A six-decade microvolt potentiometer is

junction with a 25-ohm platinum resistance thermo-

brated by the National Bureau of Standards. Low-

unsaturated standard cells are kept in insulated boxes,

arge thermal inertia thereby reducing thermal fluctua-

occasional temperature changes in the instrument

change is usually of the order of $\pm 0.5°C$). There are

rent potentiometer current supplies, one for each sec-

potentiometer. The smallest current supply of 12

res is provided by a small cadmium cell located inside

ometer, whereas the mid-section of the potentiometer

current of 1 milliampere supplied from four 1.35-volt

atteries connected in parallel. Six-volt lead-acid

upply a current of 10 milliamperes to the upper poten-

ection. This latter current supply is stabilized by

an isolated, rectified, and filtered alternating current

oss the batteries.

ble current for the platinum resistance thermometer is

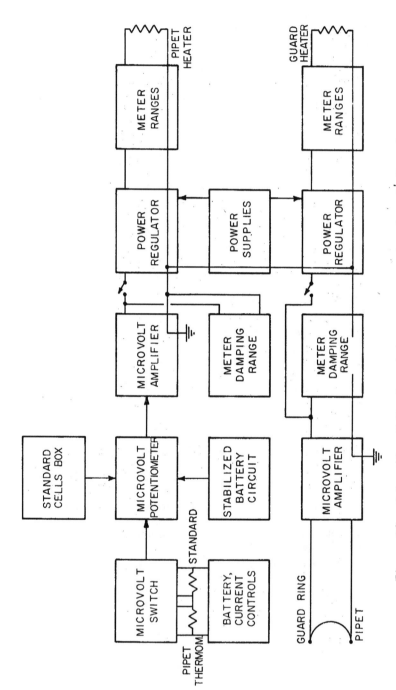

Figure 10. Schematic of the temperature measurement and control system.

Figure 11. Temperature measurement and control instruments.

obtained from four 6-volt, 200 ampere-hours, thermally insulated, and shielded batteries of low internal discharging type, connected in series. This current is adjusted when flowing through calibrated standard precision resistors. The microvolt amplifiers are coupled to 50-microamperes panel galvanometers in suitable damping and range-changing circuits. Spurious voltage effects are reduced in the thermometer and thermocouple circuits by soldering all junctions with a special cadmium-tin solder [9] resulting in negligible thermal electromotive force. All wires that connect the cryostat to the instruments are carefully shielded to avoid extraneous electrical noise and all apparatus and electrical instruments are grounded with heavy copper conductors.

The regulation and control of the pipet temperature is accomplished by balancing the cooling due to conduction down the reflux tube with Joule heating in the 500 ohm constantan wire heater. The potential across the platinum resistance thermometer, based on a thermometer current of 2 milliamperes used for all measurements, is compared to the corresponding calibrated value at the desired temperature set on the potentiometer. The difference in potential is fed through a D. C. microvolt amplifier to a power regulator [10] which adjusts the power input to the heater to maintain the pipet at the desired temperature. With an amplifier sensitivity corresponding to 0.0001 K, there is no perceptible noise.

Regulation of the guard ring temperature, identical

pipet temperature, is accomplished in a similar manner w

exception that the potential output of a differential thermoc

between the guard ring and the pipet provides the signal th

amplified and fed to the power regulator.

When standardizing, i.e., adjusting the various pote

ometer currents to give the correct voltage of the standar

the power regulator input switch is opened and the steady-

heating carefully adjusted by the power regulator control,

constant temperature is still maintained.

The meter damping and range-changing circuits con

to the galvanometers are convenient when the actual poten

differences are to be read.

Uncertainties introduced in the thermometry are pa₁

to the specification of the potentiometer of 0.01 percent of

plus 0.02 microvolts, and partly due to the calibrated valu

1-ohm standard resistor of 0.99999 ± 0.00001 ohms. In t

temperature this amounts to a maximum of 2 millidegrees

increasing to 28 millidegrees at 300 K. Uncertainty due t

calibration itself is probably less than 0.002 K.

2. Thermometry Checks

Prior to the PVT experiments with fluorine, vapor]

sures of ultra high purity (99.99 percent or better) oxyger

measured to check the accuracy of the thermometry. The observed vapor pressures, P_{obs}, are compared in Table 2 to those of the literature [4, 11], P_{lit}.

TABLE 2. Thermometry Checks from Oxygen Vapor Pressures

Temperature, K	P_{obs}, MN/m^2	P_{lit}, MN/m^2	Percent difference
124.000	1.2787	1.2788	-0.008
126.000	1.4234	1.4235	-0.007
130.000	1.7477	1.7478	-0.006
130.000	1.7479	1.7478	0.006
136.000	2.3291	2.3291	0.000
140.000	2.7863	2.7865	-0.007
144.000	3.3068	3.3063	0.015
144.000	3.3066	3.3063	0.009
146.000	3.5926	3.5918	0.022

Agreement is within the specification of the pressure measurement system, which indicates the thermometry to be of the desired absolute accuracy.

The platinum resistance thermometer used was calibrated according to the International Practical Temperature Scale of 1948 above 90 K and the NBS-1955 temperature scale below 90 K. The temperatures of Table 2 are therefore reported on this basis since the purpose of this table is comparison of data on the same temperature scale. All fluorine data, however, were converted in temperature to conform to the International Practical Temperature Scale (IPTS) of 1968 by applying the temperature scale differ-

ences of Appendix A using the Aitken interpolation method [12].
These differences for temperatures below and above 90 K were
obtained from References [13] and [14], respectively.

G.   Safety Equipment.

Due to the hazards involved in handling fluorine at high
pressures, precautions have been taken to construct and install
equipment that would prevent fluorine gas from escaping into the
laboratory in case of an accident.

As mentioned, all system components handling fluorine,
i. e., the cryostat, the gasometer, and the fluorine supply system,
are enclosed in aluminum cabinets and vented outside the laboratory by a separate ventilation system.   The 0. 32 cm thick aluminum plates should give reasonable blast protection in case of an
explosion.   In any case, the operator in the instrument room is
always separated from the high pressure fluorine system by an
explosion-proof wall.   Also, all valves subject to high pressure
fluorine are operated remotely from this room.

Should a substantial system failure accidentally develop, the
fluorine may be discarded rapidly through two charcoal reactors
installed in parallel outside the laboratory.   The reactors are
made of 9 cm diameter copper pipes lined with a layer of 1 cm
thick refractive cement since the reaction of fluorine with charcoal to form inert carbontetrafluoride, is extremely exothermic.

A vacuum system for pumping fluorine out of the PVT cell (the pipet) and the gasometer flasks was also constructed. After most of the fluorine is recovered in the booster, the remaining fraction must be removed before a new run may be started. This is done by pumping the fluorine through an alumina trap where aluminum fluoride is formed and free oxygen is released through the pumps. Two pumps, one with fluorine compatible oil for the initial pumping and another regular pump capable of producing an even lower pressure, may either pump through the trap or by-pass it.

IV. DENSITY MEASUREMENT

There are several volumes required for the density calculations. These are the volume of the gas in the gasometer system,
V_g, the volume of the pipet sample holder, V_p, and the different
nuisance volumes of the gasometer manifold, valves and capillaries. The sample density, not including the corrections of
Equation (19), is given approximately as

$$\rho = \frac{P_g}{MRT_g} \cdot \frac{V_g}{V_p} \cdot \qquad (2)$$

The subscript g refers to the gasometer system and M is the molecular weight of fluorine. Hence, the ratio V_g/V_p is an important factor affecting the density calculations.

A.   Amount of Sample Measurements.

The term $P_g V_g / MRT_g$ of Equation (2) refers to the actual
number of moles of fluorine contained in the pipet, the capillary,
and the diaphragm cell. This amount of sample may be determined through the previously discussed expansion process of the
fluorine into the appropriate glass flasks of the gasometer. To
get an accurate measure of the sample amount, it is necessary to
calibrate the glass flask volumes with utmost care.

1. Gasometer Calibration

move any dissolved air. Calculations to obtain the actual volumes
involved corrections for thermal expansion and elastic stretching
of the glass itself and for air buoyancy on the flasks and weights.
Measurements of the flask volumes were also obtained by hydrogen
gas expansion from one flask or combinations of flasks to the
other flask or flasks. The volume of the 1-liter flask as determined from the water weighings was selected as the standard
volume. Excellent agreement between volumes obtained from the
hydrogen gas expansion and the water weighings was observed for
the 2- and the 6-liter flasks, while the volume of the 21-liter
flask was taken as the mean of eight expansion determinations.
Comparison of these results are given in Table 3.

TABLE 3. Calibration of Gasometer Flask Volumes

Nominal volume, ℓ	Water weighing, cm^3	Gas expansion, cm^3	Assigned cm^3	Volume of flask metal neck, cm^3
1	988.095	(988.095)	988.095	2.432
2	2028.83	2028.80	2028.83	2.751
6	6414.71	6415.75	6415.2 ± 1.5	3.099
21		21186.5	21186.5 ± 4	3.530

Nuisance volumes of the order of a few cm^3 in capillaries, gasometer manifold, diaphragm cell, and the quartz bourdon tube were
determined through a combination of hydrogen gas expansions from
a calibrated volume of 128 cm^3 and gas expansions between individual volumes.

2. Adjustment for Sample in Capillary
and Diaphragm Cell

obtain the actual density in the pipet as described in Sec-

3. the amount of sample in the capillary and the diaphragm

he conditions of each P-T point, must be subtracted from

amount as obtained through the gasometry measurements.

rtinent data and correlations are necessary for these cal-

The temperature of the diaphragm cell, T_d,

is measured using a calibrated thermistor as

one leg of a Wheatstone bridge (the tempera-

ture inside the quartz bourdon tube is meas-

ured in the same manner.)

The temperature T_x, corresponding to any

position at a fraction x of capillary length

extending from the guard ring at the pipet

temperature, T_0, to the top of the cryostat

at the diaphragm temperature, T_d, is cal-

culated using the thermal conductivity, K,

of stainless steel [15], from

$$x = \frac{\int_{T_o}^{T_x} K \, dT}{\int_{T_o}^{T_d} K \, dT} \tag{3}$$

An equation of state is needed to calculate the compressibility factor, Z_d, at any pressure P and temperature T_d in the diaphragm cell and the compressibility factor Z_x at any pressure P and temperature T_x in the capillary. These are then used to obtain the amount of sample in these small volumes (the capillary volume, V_c, is 0.0682 cm^3 while the diaphragm cell volume V_d, including some other noxious volumes on top of the cryostat, is 0.480 cm^3). A preliminary equation of state for fluorine was obtained by fitting corresponding states data calculated from oxygen, to the equation used by Prydz and co-workers [16] for deuterium. After the new fluorine density data had been corrected in this way, it was again fitted to this equation for further refinement. The data was then corrected once more using the revised equation.

When this procedure was repeated for a

third time, no more significant changes

in densities were observed. Thus, the

amount of fluorine in the capillary is

$$N_c = \frac{P V_c \theta}{R T_0} \tag{4}$$

where the capillary factor θ is given by

$$\theta = T_0 \int_0^1 \frac{dx}{T_x Z_x} . \tag{5}$$

The number of moles of fluorine in the

diaphragm cell is given by

$$N_d = \frac{P V_d}{R T_d Z_d} . \tag{6}$$

B.   Sample Volume Measurement.

The sample volume, which is only of the order of 26 cm^3,

is the volume of the pipet, V_p. Since it is inversely proportional

to the sample density, extreme care has to be taken to arrive at

an accurate value for this volume.

1.   Determination of Pipet Volume

The volume of the normal pipet (volume at 20°C and zero

pressure) was determined by means of hydrogen gas expansions

flask resulted in a final pressure of approximately 1 atmosphere in the entire system. Initially four determinations gave a mean value of the normal pipet volume, V_0, equal to 25.881 cm^3. Later, five redeterminations resulted in a value of 25.879 cm^3 for the volume which is the value accepted for the normal volume at 20 °C for the pipet. The latter value was adopted since greater internal consistency, on the order of ± 0.003 cm^3, was obtained among the five measurements. Earlier measurements on the same pipet by Goodwin [3] yielded a volume at 20°C of 25.907 cm^3 or 0.11 percent higher than the volume of the current work. In his oxygen work Weber [4] adopted a 0.14 percent smaller volume of 25.843 cm^3.

2.   Thermal Expansion of Pipet

As noted, the normal pipet volume is a function of temperature. The volumetric thermal expansion of the pipet was obtained using known relationships [17] for the expansion coefficient of copper. Calculated values of the ratio of the normal pipet volume, V_p^0, at various temperatures to the normal pipet volume at 20°C, V_0, were fitted to a quadratic polynomial of the form

$$\lambda = \frac{V_p^0}{V_0} = 1.0 + a(T_0 - 293.15) + b(T_0 - 293.15)^2 \tag{7}$$

$a = 5.9851 \times 10^{-5}$

$b = 8.708 \times 10^{-8}$.

3. Elastic Stretching of Pipet

elationship used to calculate the elastic stretching of

e to pressure given by Goodwin [3] was modified to cor-

re closely to experimental results on similar thick-

els [18]. Thus, the corrected pipet volume at any

and temperature T_0, is given by

$$V_p = \lambda (1.0 + c[1.0 + d\ T_0]P)\ V_0 \qquad (8)$$

$\lambda = \dfrac{V_p^0}{V_0}$ as obtained from Equation (7)

$c = 2.3 \times 10^{-5}\ (MN/m^2)^{-1}$

$d = 4.35 \times 10^{-4}\ K^{-1}$.

A. Gasometry.

The number of gram moles, N, of fluorine in the particular gasometer glass flasks, the water bath manifold, the quartz bourdon tube, the capillaries, and the PVT cell during the gasometer measurements is given by

$$N = P_g/R \left[E_g V_g/T_g + (V_n + V_m)/T_m + V_{cr}/T_r + V_q/T_q \right.$$
$$\left. + V_{dg}/T_d \right]/Z_g + (V_p + V_c/2)/(T_0 Z_0)$$

where P_g is the fluorine pressure in the flasks in MN/m² as give

by

$P_g = P_q \left[1.0 + (T_q - 299.0) E_q \right]$

P_q = quartz bourdon tube pressure, MN/m²

T_q = quartz bourdon tube temperature, K

E_q = quartz bourdon tube temperature coefficient

   = 1.3×10^{-4} °C^{-1}

R = universal gas constant = 8.3143 J/mol K

E_g = glass volume pressure coefficient of flasks

   = $1.0 + (9.87 P_g - 0.825) 10^{-4}$

V_g = volume of glass flasks given in Table 3, cm³

T_g = water bath temperature = 295.00 K

V_n = volume of metal necks of flasks given in Table 3, c

V_m = volume of gasometer manifold = 6.554 cm³

T_m = gasometer manifold temperature, K

T_r = instrument room temperature, K

V_q = volume of quartz bourdon tube = 1.128 cm³

V_{dg} = volume of diaphragm cell and capillary at same temperature during gasometry measurements

= 0.767 cm³

T_d = diaphragm cell temperature, K

Z_g = compressibility factor of fluorine in flasks at T_g and P_g from Equation (27) = $1.0 - 3.626 \times 10^{-3} P_g$

V_p = pipet volume from Equation (8)

V_c = cryostat capillary volume = 0.0682 cm³

T_o = temperature of pipet, K

Z_o = fluorine compressibility factor in pipet at P_g and T_o

$$= 1.0 - 3.33030 \times 10^{-9} T_o^{-1} + 1.72408 \times 10^{-6} T_o^{-2}$$

$$- 2.22556 \times 10^{-4} T_o^{-3} \tag{11}$$

Pressure.

The fluid pressure P in MN/m² as measured in the PVT

s defined as

$$P = P_w + P_b + P_c - P_r \tag{12}$$

P = oil dead-weight gage pressure, MN/m²

g = gravity at NBS Boulder Laboratory = 9.79615 N/kg

g_s = standard gravity = 9.80665 N/kg

E_T = thermal expansion coefficient of dead-weight

gage piston = $1.5 \times 10^{-5} \, °C^{-1}$

t_w = dead-weight gage temperature, °C

P_n = nominal dead-weight gage pressure, MN/m²

P_b = barometric pressure in MN/m² = $g \, \rho_m L_m$ (14)

ρ_m = mercury density in g/cm³

= $13.5948 - 0.00245 \, t_m$ [19]   (15)

t_m = mercury temperature, °C

L_m = height of mercury column, cm

P_c = pressure exerted by the static head of fluorine

in the cryostat capillary from Equation (1).

P_r = pressure exerted by nitrogen gas column in the

instrument room = $g \, \rho_n \, L_n$ (16)

ρ_n = nitrogen density = $P/(Z_n R T_r)$ (17)

Z_n = nitrogen compressibility factor at room tempera-

ture = $1.00001 - 2.62516 \times 10^{-4} P$

$+ \, 3.24541 \times 10^{-6} \, P^2 - 2.53470 \times 10^{-9} \, P^3$ (18)

L_n = height of nitrogen column = 104 cm

C.   Density

The actual density of fluorine in the PVT cell in mol/

cm³ is obtained from

$$\rho = [N - (P/R)(V_c \theta/T_0 + V_d/(Z_d T_d))]/V_p \qquad (19)$$

where

N = total g moles of fluorine from Equation (9)

θ = capillary factor from Equation (5)

V_d = volume of diaphragm cell and including the capillary

on top of the cryostat; diaphragm in the null position

= 0.480 cm^3

Z_d = compressibility factor in diaphragm cell at P and T_d

= $1.0 + 2.58016 \times 10^{-4} P + 7.90641 \times 10^{-7} P^2$

$+ 1.59529 \times 10^{-9} P^3$. $\qquad (20)$

VI OPERATING PROCEDURE

A.   Cleaning of Fluorine System.

Before the PVT apparatus was put together, each individual
component that would be exposed to fluorine was thoroughly
cleaned. When it had been assembled it was passivated with
fluorine. The cleaning procedures were as follows:

1)   Metal parts showing visible evidence of oxidation,
      were submerged in a bath of 10 to 20 percent nitric
      acid solution to remove the oxide film. The parts
      were then rinsed with distilled water followed by
      an acetone rinse and then dried with a dry nitrogen
      or helium gas stream.

2)   Metal parts showing no visible evidence of oxida-
      tion, were washed in trichloroethylene, followed
      by an acetone rinse and then dried with dry nitro-
      gen.

3)   Polytetrafluoroethylene packings were flushed
      with trichloroethylene and acetone followed by
      a drying period under vacuum.

4)   Specific items of equipment were obtained di-
      rectly from the manufacturer ready for
      fluorine service.

B.   Passivation of Fluorine System.

The above cleaning procedures appear to have be

actory since no evidence of any reaction in the systen

'uring the subsequent passivation.   This passivation w

ed at a remote test site since experience in fluorine h

onexistent at that time.   Following is a description of

luorine passivation procedure:

1)   The system was flushed with dry nitrogen ar

then evacuated for several hours to insure r

moval of all possible moisture.

2)   The vacuum was broken with dry nitrogen ar

the system was once more evacuated.

3)   The vacuum was then slowly broken with

fluorine gas to a pressure of about 0.01 MN

and the system was held at this pressure fo

about 10 to 15 minutes.   This was followed

an increase in pressure in steps of approxi

mately 0.03 MN/m^2 until atmospheric pres

sure was reached.   The pressure was then

raised in steps of 0.2 to 1.4 MN/m^2 to som

what above the intended maximum operating

pressure and held at this pressure for

several hours.

The passivation and testing proceeded without difficulty.

C.   Charging of Storage and Recovery Cylinder.

Before proceeding with any PVT measurement it was necessary to charge the storage and recovery cylinder with fluorine
from the fluorine supply cylinder. Before charging the cylinder
all valves (numbers of valves are given in Figure 2) were initially
closed. The subsequent charging procedure is as follows:

1) Open valves 1, 2, 4, 6, 7, 8, 9, and 18 and
    evacuate the system up to valve 10.

2) Close valves 1, 2, 8, and 9.

3) Raise liquid nitrogen dewar around booster.

4) Open fluorine supply cylinder valve slightly.

5) Slightly open valve 1 and use valve 2 to regu-
    late the flow of fluorine into the valve mani-
    fold, booster and storage cylinder. Con-
    dense the fluorine in the booster.

6) Close valves 1, 2, 4, and 7. Make sure valve 6
    is open and lower the liquid nitrogen dewar
    from around the booster, allowing pressure
    to rise in the storage cylinder.

7) When the pressure reaches a maximum
    (indicating that there is no more liquid
    fluorine in the booster), close valve 6.
    Repeat steps 5, 6, and 7 until the storage
    cylinder pressure reaches a pressure of
    2 to 2·8 MN/m².

8) Close fluorine supply cylinder valve.

9) Raise liquid nitrogen dewar around
    booster again and open valves 7, 1, 2,
    and 4. Allow fluorine pressure in this
    part of the system to drop to about
    0.07 MN/m² gage pressure by conden-
    sing fluorine in the booster.

0) Close valve 7.

1) Open valve 6 and lower the liquid nitro-
    gen dewar from around the booster.

2) Close valves 1, 2, 4, and 18. Open valves
    17, 16, and 8 and evacuate this portion of
    the system.

3) Close valves 8, 16' and 17. The storage and
    recovery cylinder is now charged with fluorine.

D.   Charging of Pipet.

To charge the pipet so that a series of P-T data points may be taken at the desired psuedo-isochore, the following steps are performed:

1) Open valves 18, 15, 8, 9, 10, 11, 12, 13, and 14 and evacuate the system.

2) Close valves 8 and 10.

3) Raise the liquid nitrogen dewar around the booster and condense the fluorine in the booster.

4) Close valve 6; open valve 7 and lower the liquid nitrogen dewar from around the booster allowing the pressure to in-crease in the PVT cell. When the desired filling pressure is reached, shut valve 9.

5) Raise the liquid nitrogen dewar around the booster. When the fluorine pressure reaches a minimum, close valve 7 and open valve 6. Lower the nitrogen dewar again from around the booster.

E.   The Data Run.

1. P-T Measurements

The pipet is now ready for data taking. Pipet pressures are

at integral temperatures on the IPTS 1948 and the

temperature scales using the oil dead-weight gage. (The

ires were later converted to the IPTS 1968 temperature

previously discussed.) Data are taken at the same tem-

for isochores in the same region of the PVT surface

r the maximum desired pressure (about 20 MN/m^2) or

num temperature (300 K) is reached. Pressures are

at temperature intervals of 2 K from 54 K to 150 K.

50 K and 200 K a temperature interval of 5 K is used,

ssures are measured at every 10 K above 200 K.

2. Gasometry

noted in Chapter II, the fluorine in the pipet must be

into the gasometer to determine the number of moles

charged to the pipet originally. This is done as follows:

Close valve 15 and the appropriate valves

associated with the calibrated glass flasks

that are not to be used for the measurement.

This would involve valves 11, 12, 13 and 14.

Open valve 10 slightly and carefully allow the

3) Record the pressure in the calibrated glass
   flasks using the quartz bourdon tube gage.
   For later experimental runs the air dead-weight
   gage was used to measure this pressure, while
   the quartz bourdon tube gage served as a null
   indicator.

4) Close valve 6 and raise the liquid nitrogen
   dewar around the booster.

5) Close valve 18. Open valves 7, 8, 9,
   and 15 and condense the fluorine in the
   booster.

6) Pump on liquid nitrogen around the booster
   to reduce temperature in the booster to in-
   crease the recovery of fluorine from the
   entire system. Close valve 7 when no
   more fluorine is condensing as determined
   from the quartz bourdon tube gage.

7) Open valve 6 and lower the nitrogen dewar
   from around the booster when the solid
   nitrogen has melted.

8) Open valves 16 and 17 and evacuate the system. When a pressure of 13 N/m^2 is reached,
   the trap may be by-passed by closing valves
   16 and 17 and opening valve 18.

VII BURNETT EXPERIMENT

A.   Apparatus.

The fluorine compressibility factor in the glass flasks must be known to determine the amount of sample in the pipet from gasometer measurements. This compressibility factor, PV/RT, should be obtained at the water bath temperature of 295.00 K as a function of pressure. The gas expansion method introduced by Burnett [2] in 1936 was used for this purpose.

The Burnett apparatus, shown in Figure 12, consists of a heavy-walled AISI type 304 stainless steel bomb divided into two vessels of unspecified volumes. Vessel number 1, approximately 53 percent larger than the second vessel, is directly connected to a null pressure detector, which in turn is connected to the rest of the pressure measuring system as described previously in Subsection III. E. 1. The two vessels are connected to each other by means of a valve and capillary tubing. The apparatus, including the null pressure detector, the valves, and the tubing is submerged in a constant temperature water bath controlled at 295.00 K, the same temperature at which the gasometry measurements were taken.

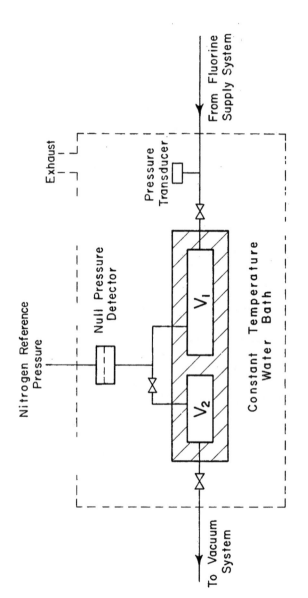

Figure 12. The Burnett apparatus.

B. Experimental Method and Analysis.

Initially, the first vessel was filled with gaseous fluorine to
the desired pressure of 2 to 3 MN/m^2, while the second vessel was
evacuated. The expansion valve was then opened and the pressure
allowed to reach equilibrium in the two volumes. Further expansions were performed until the final pressure was low enough to
allow extrapolation to zero pressure. More intermediate points
on the extrapolated curve were obtained by varying the initial
starting pressure for the expansions. Previous to the ith expansion, the fluorine gas in the first vessel is described by the following conditions:

$$(PV_1)_{i-1} = R(ZnT)_{i-1} \quad . \tag{21}$$

When equilibrium is reached after the expansion, the equation of
state becomes:

$$P_i(V_1 + V_2)_i = R(ZnT)_i \quad . \tag{22}$$

Since the temperature, T, of the water bath is constant and the
number of moles, n, also remain the same, $T_{i-1} = T_i$ and
$n_{i-1} = n_i$. Hence, division of Equation (22) by Equation (21) yields
upon rearrangement

$$\left(\frac{PN}{Z} \right)_i = \left(\frac{P}{Z} \right)_{i-1} \tag{23}$$

where

the ith expansion to the volume of the first vessel before this expansion. Thus, N_i is called the cell constant of the ith expansion. Since the pressures before and after the expansion are different, the change in the elastic stretching of the bomb results in a corresponding change in the volumes of the vessels. Consequently, the cell constant is also a function of pressure.

By extrapolating P_{i-1}/P_i of Equation (23) to zero pressure N_∞ is obtained since $Z_i = Z_{i-1} = 1$ in the limit. The value obtained for the zero pressure cell constant is 1.65335 and was determined by a least squares fit of a quadratic equation to these data. Thus, since N_∞ is known, the value of N_i at any other pressure may be obtained by using the elastic stretching formulas of Roark [20]. These equations applied to this steel bomb, relate the change of either volume from before to after the expansion as

$$\Delta V/V = 2.10 \times 10^{-5} (P_{i-1} - P_i) \tag{25}$$

where P is in MN/m^2. Thus, even at the highest operating pressures of 3 MN/m^2 this does not amount to more than a 0.004 percent change.

Substituting values of $(P/Z)_{i-1}$ for i = 2, 3, ... successively into the equation for the first expansion (i = 1), results in

for each series of expansions. The zero pressure limit

$N_2 \ldots N_i$ is $(P/Z)_0$ since $\lim_{P_i \to 0} Z_i = 1.0$. Values of

r the different expansion series may, in turn, be substi-

Equation (26), which then relates the compressibility

fluorine at the water bath temperature as a function of

. Analytically the results are represented as

$$Z(295.00 \text{ K}) = 1.0 - 3.626 \times 10^{-3} \text{ P(MN/m}^2) \qquad (27)$$

ely low pressures of a few atmospheres and are shown

ly in Figure 13.

ce no other data were available for fluorine for compari-

)ses, experimental runs were also performed using oxygen

)gen to check for accuracy. These data agreed with pub-

lues [11, 21] to within 0.005 percent. The agreement was

er near atmospheric pressures, and provided confidence

luorine compressibility data possessed similar accu-

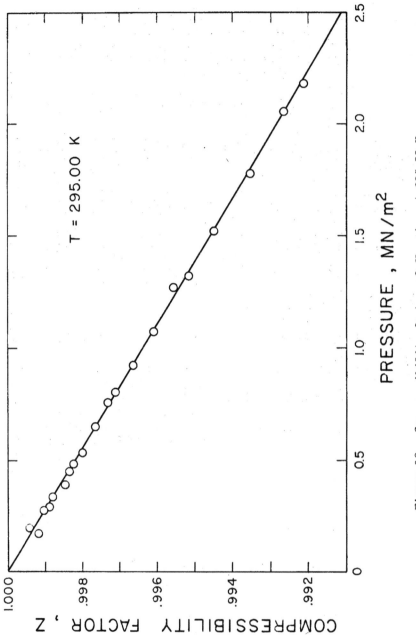

Figure 13.   Compressibility factor of fluorine at 295.00 K.

tion Densities.

tion densities derived in this study were obtained by

the experimental isochores to their intersection with

ssure equation (Equation (32)) as discussed previously

ıg the first few points of the experimental isochores

ıolynomials, the intersection of the two curves was

; an iterative procedure. The accuracy of the extra-

ds on the vapor pressure equation, the length of

to the intersection temperature, but mostly on the

slope between the experimental isochore and the vapor

·e. Thus the uncertainty of the intersection tempera-

·rom about 0.001 K for saturated liquid densities near

ıt, to more than 0.1 K for saturation densities near

:nsity where the isochores become colinear with the

·e curve. These temperature uncertainties were used

ıe relative weight factor applied to each data point

e 4) in fitting the saturation densities to the following

$$: \left| \frac{\rho - \rho_{\mathrm{c}}}{\rho_{\mathrm{c}}} \right|^3 \left[A_1 \ln \rho + \sum_{i=2}^{9} A_i \rho^{i-2} \right]. \qquad (28)$$

TABLE 4 Derived Fluorine Saturation Densities

T K	Dens., exp mol/l	Dens., calc mol/l	Pct diff	Weight
53.4811	44.86	44.862	-0.01	.000
54.541	44.690	44.696	-0.01	4109
55.572	44.541	44.533	0.02	4871
57.516	44.223	44.224	-0.00	4918
59.837	43.858	43.851	0.02	5351
59.840	43.857	43.851	0.01	4708
61.601	43.563	43.566	-0.01	4977
63.581	43.234	43.243	-0.02	4167
63.697	43.216	43.224	-0.02	4582
65.717	42.887	42.891	-0.01	5186
67.709	42.564	42.560	0.01	5552
69.842	42.206	42.202	0.01	5392
69.850	42.205	42.201	0.01	5381
71.703	41.888	41.887	0.00	5401
73.796	41.528	41.528	-0.00	5258
73.824	41.525	41.523	0.00	5292
73.834	41.521	41.522	-0.00	5290
75.381	41.252	41.254	-0.00	4748
77.421	40.894	40.897	-0.01	5100
79.813	40.470	40.474	-0.01	4970
79.811	40.470	40.474	-0.01	4984
81.837	40.112	40.110	0.01	4901
81.838	40.112	40.110	0.00	4895
83.832	39.747	39.747	0.00	4822
83.840	39.745	39.746	-0.00	4788
85.898	39.370	39.366	0.01	4710
87.855	39.005	39.000	0.01	4587
89.622	38.667	38.665	0.00	4519
91.284	38.343	38.345	-0.01	4388
93.466	37.919	37.919	0.00	4324
93.470	37.915	37.918	-0.01	4318
95.544	37.499	37.505	-0.01	3275
97.722	37.066	37.063	0.01	4089
99.972	36.596	36.596	0.00	3835
99.996	36.586	36.591	-0.01	3488
101.566	36.255	36.258	-0.01	3867
101.723	36.223	36.224	-0.00	3815
103.274	35.892	35.890	0.01	3783
105.589	35.380	35.378	0.00	3603
105.589	35.380	35.378	0.00	3573
107.242	35.006	35.004	0.00	3481
109.943	34.373	34.374	-0.00	3337
109.969	34.363	34.368	-0.01	3172
111.813	33.925	33.923	0.00	3132
113.792	33.434	33.432	0.01	3117
113.784	33.432	33.434	-0.01	3119
115.916	32.887	32.886	0.00	2975
115.916	32.887	32.886	0.00	2953
117.676	32.419	32.418	0.00	2811
119.552	31.903	31.900	0.01	2717
119.587	31.890	31.890	-0.00	2680

rine Saturation Densities--Continued

Dens., calc mol/l	Pct diff	Weight
31.287	0.00	2473
30.757	0.01	184
30.602	-0.01	2332
30.602	-0.01	2332
28.766	-0.00	1835
27.829	0.00	1604
27.829	0.00	1522
26.545	-0.04	598
25.421	0.02	487
24.166	0.02	291
23.046	-0.06	68
21.822	0.25	3
21.827	0.19	13
20.594	0.38	1
19.002	4.31	.026
18.123	3.48	.011
12.380	-3.79	.005
10.873	-0.03	.126
9.8008	-0.29	1
9.0381	-0.19	3
8.3801	0.06	4
7.7038	-0.23	8
6.8793	-0.05	45
6.6004	-0.09	29
6.0837	-0.01	119
5.4899	-0.02	76
4.7443	0.00	346
3.6611	0.08	716
2.6493	-0.02	1380
2.1181	-0.07	1557
1.5127	0.08	2412
1.1445	-0.06	2587
0.77647	0.04	2894
0.55436	-0.07	1099
0.35702	0.18	648
0.21210	-0.14	1267
0.14025	0.07	1142
0.08603	-0.06	862
0.07392	0.05	1153
0.04540	-0.05	964
0.03785	0.05	920
0.02157	0.03	366
0.00897	-0.05	216
0.00315	-0.06	422
0.00067	0.23	380
0.00057	-0.19	384

For details of weighted least-squares fitting techniques the reader
is referred to Hust and McCarty [23]. Equation (28) was recently
developed by Goodwin [24] for estimating the critical parameters of
pure fluids. These are obtained directly as the parameters resulting in the smallest variance between the experimental data and
this equation. Compared to the conventional methods this procedure
eliminates the need in this temperature range for fitting the saturated liquid and saturated vapor data separately, interpolating to
get the rectilinear diameter, and finally obtaining the intersection
of the saturation correlations with the rectilinear diameter. The
conventional method also requires better than 10^{-6} percent accuracy in the saturated liquid and vapor representation to obtain an
accuracy of 0.1 percent in the calculated saturated fluorine vapor
densities near the triple point. This requirement is estimated by
using Equation (28). Finally, accurate saturation data in the critical region are not needed as the form of Equation (28) is more
reliable in this region than the derived experimental data.

The experimental data above 120 K were fitted to Equation
(28) by the method of weighted least-squares and the critical parameters were selected as those which resulted in the smallest variance between the data and the equation. (Identical values for these
parameters were also obtained by analyzing the experimental data
using the conventional method). Only data above 120 K were used
since a better representation of the data in the critical region was

estimated critical parameters (T_c = 144. 31 K and

) had been obtained, the saturated liquid and

itted to separate functions, each being constrained

nt. This was done to obtain a better correlation

ity data near the triple point. The saturated

resentation suggested earlier [24] was modified

latic deviations between the equation and the

a were apparent. This new equation is as follows:

$$\left(\frac{\rho - \rho_c}{\rho_c} \right) = B_1 Z^{0.35} + \sum_{i=2}^{6} B_i Z^{i-1} \qquad (29)$$

$Z = 1 - T/T_c$.

resulting from this weighted least-squares fit of

lid densities are

$B_1 = 1. 81881076$

The number of significant figures given for these coefficients and for those of subsequent equations is not justified on the basis of the uncertainty of the data, but is presented to enable duplication of the calculated values. Applied weight factors and the deviations between the experimental and calculated densities are given in the upper part of Table 4.

Similarly, all vapor densities were represented by the equation suggested by Goodwin [24]

$$\ln \left(\frac{\rho}{\rho_c} \right) = C_1 \left(\frac{Z}{Z-1} \right) + C_2 Z^{0.35} + \sum_{i=3}^{7} C_i Z^{i-2} . \tag{30}$$

Calculated saturated vapor data below 80 K were also fitted to insure a representative equation that gives saturation densities approaching the ideal gas value at the triple point (vapor pressure = 252 N/m^2). These data were based on the vapor pressure equation and an equation of state for fluorine [25] which represents the PVT surface in this region to within the uncertainty of the data. The following coefficients are obtained from this fit:

$$C_1 = 4.85547085$$

$$C_2 = -1.96015519$$

$$C_3 = -1.88066900 \times 10^{-1}$$

$$C_4 = 6.21165939$$

$$C_5 = -2.29600897 \times 10^1$$

s, however, the thermodynamic properties were

he saturated vapor densities below 6.0

K) on the virial equation (34). Experi-

144.31 K in Table 4 were used in Equation (28)

tical parameters but were not used in Equations

vious reasons (Z becomes negative). The re-

:s in all equations (as discussed earlier) were

1968 temperature scale.

quid density data from other sources [26, 27,

re been compared to Equation (29). All data

:rever possible, for temperature scale differ- ·

. were not, however, included in the final fit of

y deviate from the new data by as much as 28

(e.g. Kanda's density data [26] are too low by

ilner et al. [27] report one density measure-

. is 1.0 percent too low (their reported saturated

xygen is low by about 1.7 percent). Saturated

)orted graphically by Dunn and Millikan [28] are

vhich Elverum and Doescher [29] of the same

ported numerical values. These data appear to

nt too high in density. The density data by

White, Hu and Johnston [30] are the smoothest of the old data but
are too high by about 0.4 percent. Reported measurements by
Jarry and Miller [31] agree well with those of Reference [30], but
they exhibit larger internal scatter (0.3 percent). Early measurements of the saturated liquid density at the normal boiling point
[32, 33, 34] were also compared in this study. These data points
deviate considerably from the new data due to large uncertainties
in the temperature measurements as well as the density determinations.

No published saturated vapor density data exist but calculated values based on the Berthelot equation of state by Fricke [35]
deviate from Equation (30) by as much as + 10 percent in density
at 140 K.

B.   The Critical Point.

The critical temperature and density of fluorine estimated
from Equation (28) are given in Table 5. Also given is the critical
pressure obtained from the vapor pressure equation (see Equation
(32)). Critical parameters from other sources are also compared
to the new values in this table.

Fricke [35] calculated the critical density from the
Berthelot equation of state, while Baker and Adamson [36] used
the critical temperature from Reference [37] and obtained the
critical density by generalized properties techniques. Goodwin [24]

T_c, IPTS 1968 K	ρ_c mol/l	P_c MN/m^2	Reference
143.9	15.1		[24]
	12.4		[35]
	15.0		[36]
144		5.57	[37]
144.31 ± 0.05	15.10 ± 0.04*	5.215	This research

* The estimates of the uncertainty in these critical parameters were based on the experience gained through the analysis of the data since no direct statistical measure is obtained from this iterative fitting procedure.

C.   Melting Line Determination.

Fourteen pressure-temperature data points were measured along the fluorine melting line during experimental runs no. 121 and 122 as reported earlier [38]. Measurements were made at 0.1 K intervals from 1.3 to 13.7 MN/m^2. These data, given in Table 6, were fitted to the Simon melting equation

$$P = P_t + P_o \left[(T/T_t)^c - 1 \right] \tag{31}$$

where the subscript t refers to the triple point. By varying the values of c and T_t a best fit was obtained with the following

$$c = 2.1845$$

$$T_t = 53.4811 \text{ K.}$$

Only two measurements on the normal melting point and none at higher pressures have been made prior to this study. Hu, White, and Johnston's [39] value of 53.53 K agrees fairly well with the new triple point temperature, whereas Kanda's [40] measurement of 55.2 K appears to be too high. The reason for the differences may be due to impurities in the fluorine sample of the earlier data or temperature scale differences.

BLE 6 Fit of the Simon Equation to Melting Curve Data

P exp MN/m^2	P calc MN/m^2	Diff MN/m^2
1. 3141	1. 3136	0. 0005
2. 3350	2. 3371	-0. 0021
3. 3653	3. 3628	0. 0025
4. 3943	4. 3907	0. 0036
4. 3901	4. 3907	-0. 0006
5. 4182	5. 4208	-0. 0026
6. 4487	6. 4539	-0. 0052
7. 4857	7. 4889	-0. 0032
8. 5322	8. 5258	0. 0064
9. 5650	9. 5649	0. 0001
10. 6060	10. 6060	-0. 0000
11. 6500	11. 6493	0. 0007
12. 6979	12. 6949	0. 0030
13. 7392	13. 7427	-0. 0035

IX VAPOR PRESSURE DETERMINATION

Fluorine vapor pressure data are reported in the literature by several experimenters [37, 39, 41, 42]. However, these
are inconsistent measurements covering the range from the triple
point to just a few degrees above the normal boiling point. To
insure accuracy in the subsequent thermodynamic properties calculations it was necessary to obtain consistent vapor pressure
measurements extending from the triple point to the critical point.
This was done using the PVT pressure measuring system together with the gasometer pressure gages as shown schematically
in Figure 14 [43].

A.   Experimental Approach.

Pressures were measured by three different procedures
in overlapping ranges from the triple point to the critical point.
Pressures below about 0.2 MN/m^2 were measured directly using
the fused quartz bourdon tube pressure gage previously calibrated
to the air dead-weight gage. Pressures above 0.003 MN/m^2 to
about 4 MN/m^2 were measured by referencing the fluorine to .
nitrogen pressures derived directly from the air piston dead-
weight gage using a diaphragm type differential pressure transducer. Above 0.13 MN/m^2, pressures were referenced to the oil
dead-weight gage using two differential pressure transducers with
an intermediate nitrogen system separating the oil and the fluorine

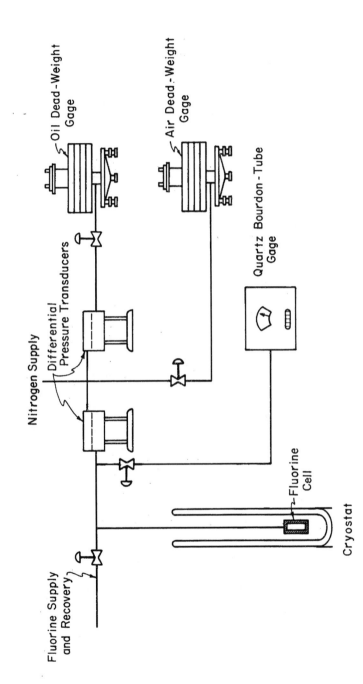

Figure 14. Pressure measuring system for vapor pressure determinations.

for safety purposes.

Each vapor pressure reading was made at least twice at a given temperature with some fluorine being removed from the cell between readings. Identical pressure readings verified the fact that two-phase conditions existed in the cell during a measurement.

B.   Correlation of Vapor Pressure Data.

The fluorine vapor pressure data in Table 13 obtained as a result of this work were represented by the following equation:

$$ \ln \left(\frac{P}{P_t} \right) = A_1 X + A_2 X^2 + A_3 X^3 + A_4 X \left(1 - X \right)^{A_5} \quad (32) $$

where

$$ X = (1 - T_t/T)/(1 - T_t/T_c) $$

$$ P_t = 2.52 \times 10^{-4} \ MN/m^2 $$

$$ T_t = 53.4811 \ K $$

$$ T_c = 144.31 \ K $$

The subscripts t and c refer to the triple point and the critical point, respectively. A trial and error method was used to select the value of the triple point pressure which resulted in the smallest sum of the squared deviations between the data and the corresponding calculated values. Changes in this pressure of the order of 0.5 percent were directly reflected in larger deviations all the way to the critical point. The obtained value agrees well th the pressure calculated by Ziegler and Mullins

(32) was recently developed by Goodwin [45] through

e study of vapor pressure equations. It is non-

ritical point, i.e., the curvature is infinite, and

le slope dP/dT over the entire liquid range, a

t the polynomial type equations do not always meet.

efficients, A_i's, of Equation (32) were determined

es fit of the experimental data in Table 13. An

ure was adopted such that the chosen value for A_5

eighted sum of the squared deviations. However,

must be between 1.0 and 2.0 to give infinite curva-

al point. The selected coefficients for the vapor

n with P in MN/m^2 and T in degrees K, are

$A_1 = 7.89592346$

$A_2 = 3.38765063$

$A_3 = -1.34590196$

formula, due to errors in temperature. Hence,

$$W = \frac{1}{\sigma_P{}^2 + \left(\frac{\partial P}{\partial T.} \sigma_T \right)^2}$$

where $\sigma_P{}^2$ and $\sigma_T{}^2$ are the respective variances in pressure and temperature.

C.   Discussion.

The experimental data are compared to the vapor pressure equation in Figure 15. Pressure deviations are defined as $(P_{exp} - P_{calc}) \times 100/P_{exp}$, where P_{exp} is the experimental pressure and P_{calc} is the value calculated from Equation (32). Data by other experimenters [37, 39, 41] were adjusted in temperature whenever possible to conform with the IPTS 1968 temperature scale and compared to the new data in Figure 15. Rigorous calculated values [44] based on thermodynamic data such as latent heats due to phase changes, heat capacity and mean molal volume of the liquid, etc. and fluorine molecular constants [46] are also compared in this figure. However, the data from these sources were not included in the final fit of the vapor pressure equation since they deviate from the new data by as much as 16 percent. These large deviations are partly due to problems involved in the early fluorine vapor pressure measurements, which involved reaction of fluorine with mercury in the glass

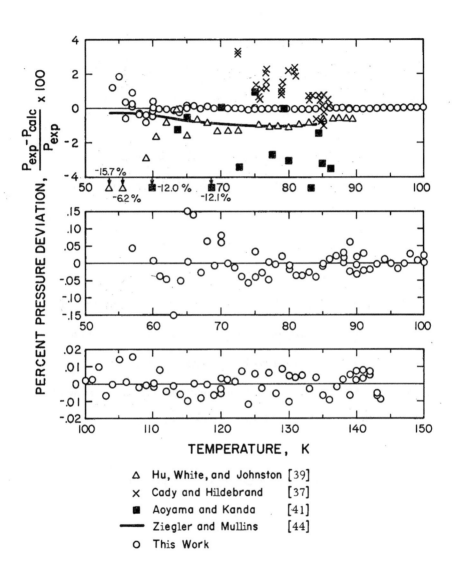

Figure 15. Deviations of vapor pressure data from Equation(32).

equalizing manometer [37] and the difficulty in obtaining a pure
fluorine sample. The data by Claussen [42] were not considered
as they were given in terms of a vapor pressure equation. Deviations of the data from this equation were reported to be as much
as 2.5 percent.

No previous measurements of fluorine vapor pressure
exist above 89.4 K, but values given by Landau and Rosen [47]
obtained by extrapolating the data of Cady and Hildebrand [37] to
higher temperatures, appear to be in error by as much as - 15
percent.

By interpolating the data in Table 7 directly, a normal
boiling point temperature of $84.95_0 \pm 0.003$ K on IPTS 1968 was
obtained. This agrees well with published values of 85.03 K [39]
and 84.95 K [37], corrected for temperature scale differences.

Calculated vapor pressures [44] are based on the normal
boiling point of [39] and hence agree well with that data in this
region, but they deviate by as much as + 15 percent from [39] at
the triple point.

l liquid and vapor regions are given in Table 14. The

of this table labeled "IDENT" contains the identification

of each data point. The last two digits represent the

imber while the remaining digits are the number of the

experimental isochore. Entries in runs 120 and 121 which

no value for the density are pressures measured on the

curve.

Some attempts were made to fit this data with an equation

used successfully for other cryogenic fluids [25]. How-

his equation was not capable of representing the data to

ts precision, a fact particularily evident in the critical

Therefore, to obtain the best possible representation of

data, it was decided to divide the PVT surface into two

as pictured in Figure 16. The data in each region were

ted separately as discussed below. No other published

phase fluorine PVT data were available for comparison to

data.

Low Density Region.

Experimental PVT data of densities less than 0.22

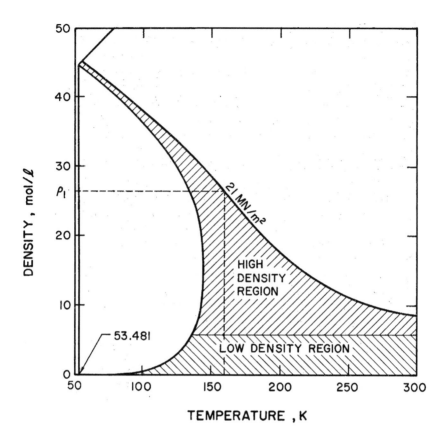

Figure 16. Density-temperature diagram showing the divided surface used for computations of thermodynamic properties.

;ing the air dead-weight gage while the density was de-

₂ating the PVT cell to 175 K using the known pressure-

ationship for that isotherm.   These data together with

xperimental PVT data below 6.0 mol/l in density were

d by the truncated virial equation

$$P = RT \left[\rho + B(T)\rho^2 + C(T)\rho^3 \right] . \tag{34}$$

l and third virial coefficients were expressed as a

temperature by

$$B(T) = \sum_{i=1}^{5} B_i T^{(1-i)/4} \tag{34a}$$

$$C(T) = \sum_{i=1}^{6} C_i T^{(1-i)/2} , \tag{34b}$$

ly. An iterative method was used to optimize both the

coefficients of $B(T)$ and $C(T)$ and their temperature

he decision to use 6.0 mol/l as a maximum density

₁e virial surface was based on the results obtained by

₂xperimental data to Equation (34).   By varying the

₂xperimental density included in the fit of this equation,

hibited definite systematic deviations from the equation

as well as temperature for all maximum densities

greater than 6.0 mol/l. Equation (34) was based on the 329 PVT

data points in the virial region. Further, 26 experimental single-

phase specific heat (C_v) data points [48] between 6.5 and 6.8 mol/l

at temperatures from 140 K to 300 K were compared to extrapolated

C_v values calculated from the virial equation. The reason for this

was to ascertain that the apparent deviations were fairly constant

and not a function of temperature. The coefficients of Equations

(34a) and (34b) and their significance level obtained through a

weighted least-squares fit of all the PVT data in this region are

given in Table 7. Weight factors [23] used in the fitting of Equa-

tion (34) were similar to those used for the vapor pressure equation

(32), except variances in density were also considered. PVT data

below 100 K were weighted much lower than above this temperature

as the uncertainty of the measurements in this region increases

rapidly with decreasing density. The weighted standard deviation

in density between the virial surface and all of the 329 PVT data

points is 0.041 percent. For consistency, Equation (34) was con-

strained to give the same saturation temperature (137.327 K) for

the 6.0 mol/l isochore as is obtained from Equation (30).

The second and third virial coefficients, calculated from

Equations (34a) and (34b), are given in Table 8. Also given in

this table are the quantities δB and δC which are the respective

estimated uncertainties of the coefficients based on the experience

nt	Least squares estimate of coefficient	Standard deviation of coefficient	Significance level in percent*
	-4.43719523	1.10	$99.9+$
	6.88646977×10	1.61×10	$99.9+$
	-4.00652537×10^2	8.85×10	$99.9+$
	1.04730534×10^3	2.15×10^2	$99.9+$
	-1.05492603×10^3	1.96×10^2	$99.9+$
	$3.97288149 \times 10^{-1}$	6.38×10^{-2}	$99.9+$
	-2.80769183×10	4.33	$99.9+$
	7.95698766×10^2	1.18×10^2	$99.9+$
	-1.12867697×10^4	1.59×10^3	$99.9+$
	8.01450388×10^4	1.08×10^4	$99.9+$
	-2.27594177×10^5	2.93×10^4	$99.9+$

parameters are significant at the level indicated when

g the standard F-test.

TABLE 8 Second and Third Virial Coefficients of F

(δB and δC are estimated uncertainties)

T K	B	δB x 10^3	C	
		1/mol		(1/m
80	−0.2396	40	−0.022557	
85	−0.2132		−0.013548	
90	−0.1910		−0.007748	
95	−0.1722		−0.004016	
100	−0.1561	10	−0.001624	
105	−0.1422		−0.000106	
110	−0.1301		0.000838	
115	−0.1194		0.001409	
120	−0.1100		0.001736	
125	−0.1017	2	0.001905	
130	−0.0942		0.001973	
135	−0.0875		0.001979	
140	−0.0815		0.001947	
145	−0.0759		0.001893	
150	−0.0709	.3	0.001828	
155	−0.0663		0.001758	
160	−0.0621		0.001689	
165	−0.0582		0.001621	
170	−0.0546		0.001557	
175	−0.0512		0.001498	
180	−0.0481		0.001442	
185	−0.0452		0.001391	
190	−0.0425		0.001344	
195	−0.0400		0.001301	
200	−0.0376	.3	0.001261	
205	−0.0354		0.001224	
210	−0.0332		0.001190	
215	−0.0313		0.001159	
220	−0.0294		0.001130	
225	−0.0276		0.001104	
230	−0.0259		0.001080	
235	−0.0243		0.001057	
240	−0.0228		0.001037	
245	−0.0214		0.001019	
250	−0.0200	.3	0.001003	
255	−0.0187		0.000988	
260	−0.0175		0.000976	
265	−0.0163		0.000966	
270	−0.0152		0.000957	
275	−0.0141		0.000951	
280	−0.0131		0.000947	
285	−0.0122		0.000946	
290	−0.0112		0.000946	
295	−0.0104		0.000949	
300	−0.0095	.3	0.000955	

ient with values obtained by White, Hu, and Johnston

in Figure 17. In the calculation of their data these

cted the contribution of the third and higher order

ients, which is the probable reason for their larger

es, especially at lower temperatures. Temperature

ices could account for part of the deviations. Since

Reference [49] do not represent just the second

ient, further comparisons with Equation (34a) were

erting values from oxygen [4] and argon [50] using

iding states approach based on critical parameters

$$B_{converted} = B(X) \; \frac{\rho_c \, (X)}{\rho_c \, (F_2)}$$

$$T_{converted} = T(X) \; \frac{T_c \, (F_2)}{T_c \, (X)} \; ,$$

presents either oxygen or argon. The agreement is

be seen from Figure 17. In fact, the correspondence

s calculated from the second virial coefficient of

quation (34a) is better than one may justifiably expect

ciple of corresponding states.

ublished values for the third virial coefficient of

available for comparison to Equation (34b).

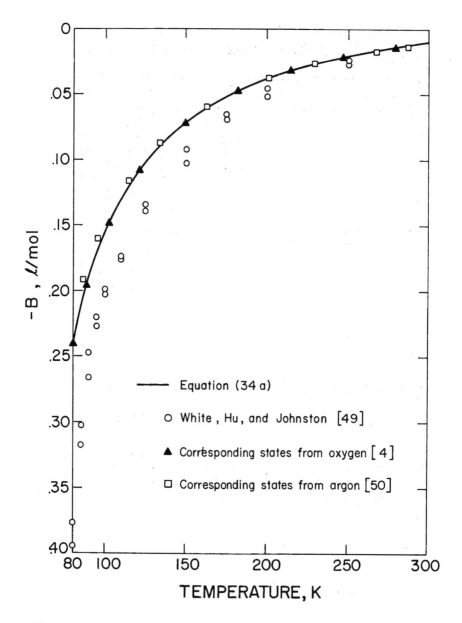

Figure 17. Comparisons of second virial coefficients of fluorine.

1. Isotherms

The high density region (see Figure 16) bounded by the mol/l isochore, the vapor-liquid coexistence boundary, melting curve, and the 21 MN/m^2 isobar was represented by isotherm polynomials of the form

$$P = RT\rho + \sum_{i=1}^{n} A_i \rho^{i-1} , \qquad (35)$$

re $A_1 = A_2 = 0$ for $T > T_c$ or $\rho < \rho_c$.

ι points for a fictitious liquid isotherm at 53.4 K, used for the ιose of extrapolating to the melting curve, were obtained from xtrapolation of isobars calculated from the isotherms between : and 64 K. A few points on the 54 K isotherm were also obed in this manner. The density of the liquid in equilibrium the solid derived from the mentioned isobars, was expressed function of either temperature or pressure by

$$\rho_{melt\ L} = \rho_t + 0.208\ (T - T_t),\ mol/l \qquad (36)$$

$$\rho_{melt\ L} = \rho_t + 0.020\ P(MN/m^2). \qquad (37)$$

lue of 44.86$_2$ mol/l was used for the liquid density, ρ_t, e triple-point.

No more than 3 coefficients were used for the isotherms below 60 K since these consisted of only 4 P-ρ data points each. As many as 9 coefficients were necessary for the 146 K and 148 K isotherms (slightly above the critical temperature) to adequately represent the data. From 4 to 7 coefficients were needed to represent all of the remaining isotherms. The isotherms in the compressed liquid and gaseous phases below the critical temperature were constrained to the corresponding saturation density of Equations (29) and (30), respectively. Above 138 K the isotherms were also fitted to the low density data in the virial region to provide a better match of the surfaces around the 6.0 mol/l isochore.

The standard deviation of the 381 experimental data points in the compressed liquid was only 0.005 percent in density while the 540 compressed vapor data points exhibit a larger deviation of 0.018 percent. The latter value is higher due mainly to the contribution from the isotherms just above the critical temperature where the deviations are much larger than over the rest of the surface. However, this is what one normally would expect.

2. Isochores

Most calculations of thermodynamic properties require knowledge of the quantities $(\partial P/\partial T)_\rho$ and $(\partial^2 P/\partial T^2)_\rho$. In the low density region these derivatives may be calculated directly from

(34) while for densities abóve 6. 0 mól/l it was necessary

ate pressures at even density increments. These isochores

n represented using functions of the·form

$$P = \sum_{i=1}^{n} A_i \, T^{(3-2i)} \tag{38}$$

ncrements of 0. 5 mol/l were used for the 80 isochores

5. 5 and 45. 0 mol/l. The number of terms of Equation (38)

om 7 at the lowest density to 3 at the 45. 0 mol/l isochòre.

fit of the isotherms was directly carried over in the

on of these isochores.

XI DERIVED THERMODYNAMIC PROPERTIES

Additional data to the PVT measurements are necessary in the calculations of derived thermodynamic properties. These are mainly thermodynamic functions of the ideal gas; however, specific heat data of the single-phases and the saturated liquid also help to improve the accuracy of the calculations.

A.   Thermodynamic Functions of Ideal Gas.

Since new measurements of the dissociation energy of fluorine [51] recently became available, it was decided to recompute the ideal gas thermodynamic functions of this fluid. Maximum changes in the functions compared to earlier values [46] were less than 0.1 percent for temperatures of 300 K and below. These thermodynamic functions were calculated according to the statistical mechanical treatment of Mayer and Mayer [52] for diatomic molecules. The relationships based on their approach include first order corrections to the harmonic oscillator-rigid rotator for: vibrational anharmonicity, rotational stretching, and rotation-vibration interaction. No attempt will be made here to define all different parameters. The conventional nomenclature for these variables may be obtained from standard texts [52, 53]. The correlations are as follows:

Translational and rotational contributions:

$$-(F^\circ - H_0^\circ)/RT = 1.5 \ln M + 2.5 \ln T - 3.6644173$$
$$-\ln \sigma + \sigma/3 + \sigma^2/90 \tag{39}$$

$$(H^\circ - H_0^\circ)/RT = 3.5 - \sigma/3 - \sigma^2/45 \tag{40}$$

$$C_p^\circ/R = 3.5 + \sigma^2/45 \tag{41}$$

where $\sigma = (B_e - \alpha_e/2) \, hc/kT = 1.27028/T.$ \tag{42}

Vibrational contributions:

$$-(F^\circ - H_0^\circ)/RT = -\ln (1 - e^{-u}) \tag{43}$$

$$(H^\circ - H_0^\circ)/RT = ue^{-u}/(1 - e^{-u}) \tag{44}$$

$$C_p^\circ/R = u^2 e^{-u}/(1 - e^{-u})^2 \tag{45}$$

where $u = (\omega_e - 2\chi_e \omega_e) \, hc/kT = 1283.30/T.$ \tag{46}

Anharmonicity contributions:

$$-(F^\circ - H_0^\circ)/RT = 8\gamma^2/\sigma + \delta e^{-u}/(1 - e^{-u})$$
$$+ 2X \, ue^{-2u}/(1 - e^{-u})^2 \tag{47}$$

$$(H^\circ - H_0^\circ)/RT = 8\gamma^2/\sigma + \delta ue^{-u}/(1 - e^{-u})^2 + 4X \, u^2 e^{-2u}/(1 - e^{-u})^3$$
$$- 2X \, ue^{-2u}/(1 - e^{-u})^2 \tag{48}$$

$$C_p^\circ/R = 16\gamma^2/\sigma - \delta u^2 e^{-u}/(1 - e^{-u})^2 + 2\delta u^2 e^{-u}/(1 - e^{-u})^3$$
$$- 8X u^2 e^{-2u}/(1 - e^{-u})^3 - 4X u^3 e^{-2u}/(1 - e^{-u})^3$$
$$+ 12X u^3 e^{-2u}/(1 - e^{-u})^4 \tag{49}$$

where

$$\delta = \alpha_e / B_e = 0.02107 \tag{50}$$

$$8\gamma^2 = 2D_e / B_e = 7.5095 \times 10^{-6} \tag{51}$$

$$X = \omega_e \chi_e / \omega_e = 0.02064. \tag{52}$$

Finally, values of entropy were computed from the following relationship

$$S^\circ / R = (H^\circ - H_0^\circ)/RT - (F^\circ - H_0^\circ)/RT. \tag{53}$$

The variables σ, u, δ, γ, and X were calculated from the new measurements of the dissociation energy of fluorine [51] and the Raman spectrum measurements of Andrychuk [54].

Table 9 list some representative values of the ideal gas functions used in the calculations of the thermodynamic properties of fluorine. Further checks on the ideal gas enthalpy, H°, and entropy, S°, were obtained by numerically integrating the ideal gas specific heat at constant pressure, C_p° (Equations (41, 45, 49)) and C_p° / T as functions of temperature, respectively. All deviations from the values given in Table 9 were less than 0.001 percent. H_0° was taken to be zero in all calculations.

B.   Calculational Methods.

Obviously no integration with respect to density of the thermodynamic equations was possible until a final density was determined for a given temperature and pressure. This was done

3.50163	3.49639	20.47877
3.50250	3.49690	20.81255
3.50407	3.49743	21.11736
3.50668	3.49803	21.39793
3.51062	3.49878	21.65794
3.51617	3.49974	21.90033
3.52350	3.50099	22.12748
3.53271	3.50258	22.34136
3.54381	3.50455	22.54359
3.55675	3.50695	22.73553
3.57140	3.50980	22.91833
3.58762	3.51311	23.09297
3.60520	3.51689	23.26027
3.62396	3.52114	23.42093
3.64370	3.52583	23.57558
3.66420	3.53095	23.72474
3.68530	3.53648	23.86886
3.70681	3.54239	24.00835
3.72857	3.54865	24.14355
3.75045	3.55523	24.27477
3.77233	3.56210	24.40228
3.79410	3.56924	24.52633

by an iterative process of the virial equation for densities below

6.0 mol/l and of the isotherms for densities above 6.0 mol/l.

Once the density had been established, the calculations were per-

formed using the following equations.

For densities less than critical at T less than

T_2 (= 160 K) and for all densities for $T \geq T_2$, the thermodynamic

properties were calculated as follows:

$$H(T, \rho) = H^\circ(T) + \frac{P}{\rho} - RT$$

$$+ \int_0^\rho \left[\frac{P}{\rho^2} - \frac{T}{\rho^2} \left(\frac{\partial P}{\partial T} \right)_\rho \right] d\rho \, , \qquad (54)$$

$$S(T, \rho) = S^\circ(T) - R \ln \left(\frac{RT\rho}{P_0} \right)$$

$$+ \int_0^\rho \left[\frac{R}{\rho} - \frac{1}{\rho^2} \left(\frac{\partial P}{\partial T} \right)_\rho \right] d\rho \, , \qquad (55)$$

where P_0 = 1 atm;

$$C_v (T, \rho) = C_v^\circ (T) - T \int_0^\rho \frac{1}{\rho^2} \left(\frac{\partial^2 P}{\partial T^2} \right)_\rho d\rho \, . \qquad (56)$$

The ideal gas properties for enthalpy, entropy, and specific heat,

H°, S°, and $C_v^\circ (= C_p^\circ - R)$, respectively were obtained from

Equations (39) through (53). All integrations in the high density

region of Figure 16 were carried out numerically using the

trapezoidal rule. For the development of these and the following equations the reader is referred to References [55, 56].

For densities greater than the critical between tempera-
atures $T_1 = 135$ K and $T_2 = 160$ K the starting points for the calculations are taken as those calculated from the above equations at
T_2 and a density $\rho_1 = 26.4276$ mol/l. The specific heats along this
isochore, given by Equation (69), were used to compute the changes
in properties as a function of temperature:

$$H(T, \rho_1) = U(T_2, \rho_1) + \int_{T_2}^{T} C_v(T, \rho_1)\, dT + P(T, \rho_1)/\rho_1 , \qquad (57)$$

$$S(T, \rho_1) = S(T_2, \rho_1) + \int_{T_2}^{T} [C_v(T, \rho_1)/T]\, dT \qquad (58)$$

where the internal energy was obtained from

$$U(T_2, \rho_1) = H(T_2, \rho_1) - P(T_2, \rho_1)/\rho_1 . \qquad (59)$$

Further computations were made along isotherms using

$$H(T, \rho) = H(T, \rho_1) + \int_{\rho_1}^{\rho} [P - T\,(\partial P/\partial T)_\rho]/\rho^2\, d\rho$$

$$+ P(T, \rho)/\rho - P(T, \rho_1)/\rho_1 , \qquad (60)$$

$$S(T, \rho) = S(T, \rho_1) - \int_{\rho_1}^{\rho} [(\partial P/\partial T)_\rho/\rho^2]\, d\rho , \qquad (61)$$

and

$$C_v'(T, \rho) = C_v(T, \rho_1) - T \int_{\rho_1}^{\rho} [(\partial^2 P/\partial T^2)_\rho/\rho^2] \, d\rho. \tag{62}$$

For the compressed liquid at temperatures less than T_1 (= 135 K) use was made of the properties of saturated liquid at (T_1, ρ_1) as a starting point for the computations. Changes in properties along the saturated liquid curve were calculated from the following equations:

$$H(T, \rho_{sat\,L}) = H(T_1, \rho_{sat\,L}) + \int_{T_1}^{T} C_{sat} \, dT$$

$$+ \int_{P_{sat}(T_1)}^{P_{sat}(T)} (1/\rho_{sat\,L}) \, dP, \tag{63}$$

$$S(T, \rho_{sat\,L}) = S(T_1, \rho_{sat\,L}) + \int_{T_1}^{T} (C_{sat}/T) \, dT, \tag{64}$$

and

$$C_v(T, \rho_{sat\,L}) = C_{sat}(T) + \frac{T}{\rho_{sat\,L}^2} \left(\frac{d\rho_{sat\,L}}{dT} \right) \left(\frac{\partial P}{\partial T} \right)_v. \tag{65}$$

Since the integral of Equation (64) cannot be solved analytically and since $(1/\rho_{sat\,L})$ was not correlated as a function of the vapor pressure, these integrals were obtained numerically. $C_{sat}(T)$ is the saturated liquid specific heat as discussed in the following section and given by Equation (68). Further computa-

$$U(T, \rho) = H(T, \rho) - P(T, \rho)/\rho, \qquad (66)$$

$$C_p(T, \rho) = C_v(T, \rho) + (T/\rho^2)(\partial P/\partial T)_\rho^2 /(\partial P/\partial \rho)_T . \qquad (67)$$

Specific Heat Data Used

Specific heat measurements of fluorine [48, 57], recently

pleted in this laboratory, were used to improve the accuracy

e thermodynamic properties and make the thermodynamic

ulations more consistent. In the calculations for the com-

sed liquid and the compressed gas below 160 K use was made

e specific heat of saturated liquid [57], G_{sat} (this quantity

not include the specific heat of the corresponding saturated

r), expressed in the following form

$$C_{sat}(T) = A_1 (T_c - T)^{A_2} + \sum_{i=3}^{7} A_i T^{i-3} . \qquad (68)$$

coefficients for this equation are given in Table 10. The

-mean square deviation of the 34 C_{sat} data points between the

e and the critical point is 0.11 percent.

region according to Equation (62). This C_v isochore intersects the liquid saturation boundary at a temperature of 135.000 K (T_1) and the 160 K (T_2) isotherm (T_2 = 160.010 K on IPTS 1968) at a pressure of 20.8 MN/m^2. Analytically, these specific heat measurements at the ρ_1 - isochore were represented by

$$C_v(\rho_1) = \sum_{i=1}^{4} A_i T^{i-3} \qquad (69)$$

where the coefficients are given in Table 10. To obtain even more consistent results, Equation (69) was constrained at $C_v(T_1, \rho_1)$ = 28.282 J/mol as calculated by the following expression

$$C_v(T_1, \rho_1) = G_{sat}(T_1) + \frac{T_1}{\rho_1^2} \left[\left(\frac{d\,\rho_{sat\,L}}{dT} \right) \left(\frac{\partial P}{\partial T} \right)_{\rho_1} \right]_{T_1}. \qquad (70)$$

Equation (69) was based on 20 adjusted (to ρ_1) experimental C_v points which exhibit a root-mean-square deviation of 0.09 percent when the equation is constrained. Prior to constraining Equation (69) the value of $C_v(T_1, \rho_1)$ calculated from this equation (Eq. (69)) was about 0.2 percent less than that obtained from Equation (70).

ıis is certainly within the accuracy of C_{sat} (135 K). Further,

e quantity (d $\rho_{sat\,L}/dT$) is the temperature dependency of the

TABLE 10   Coefficients for Equations (68) and (69)

Coefficient	Equation (68)	Equation (69)
A_1	2.05286929 x 10^2	5.06916112 x 10^6
A_2	-0.593	-9.49753733 x 10^4
A_3	-4.29836188	6.25979166 x 10^2
A_4	2.08808261	-1.27643818
A_5	-3.45371635 x 10^{-2}	
A_6	2.40274037 x 10^{-4}	
A_7	-6.13321958 x 10^{-7}	---

.turated liquid density (Equation (29)) and $(\partial P/\partial T)_{\rho_1}$ is the ter-

inal slope of the ρ_1-isochore at the vapor-liquid coexistence

undary. These slopes cannot be estimated with confidence by

trapolating the PVT data because the isochores in this terminal

gion may have maximum curvature. Considering that G_{sat} is

out $3C_v$ at T_1 (135 K) and that the bases for the thermodynamic

operties calculation are Equations (68) and (65) below T_1 while

ıuation (69) is used as a starting point above this temperature, are given in Table 11. This curve is defined as the locus of points where the Joule-Thomson coefficient is equal to zero. This condition may be described in classical thermodynamics by the relationship

$$- \frac{T}{\rho^2} \left(\frac{\partial \rho}{\partial T} \right)_p = \frac{1}{\rho} \tag{71}$$

or more conveniently

$$T \left(\frac{\partial P}{\partial T} \right)_\rho = \rho \left(\frac{\partial P}{\partial \rho} \right)_T . \tag{72}$$

Equation (72) may be solved by an iterative technique using the isotherm derivative, $(\partial P/\partial \rho)_T$, and the isochore derivative, $(\partial P/\partial T)_\rho$, obtained from Equations (35) and (38), respectively. The computed pressure errors in Table 11 are based on an assumed uncertainty of 1 percent in these derivatives.

Tables 15 and 16 present the results of all thermodynamic property calculations on the vapor-liquid saturation boundary and on selected isobars, respectively. The velocity of sound, W_s, was calculated from the following relationship

$$W_s = [(C_p/C_v) (\partial P/\partial \rho)_T]^{1/2} . \tag{73}$$

The variation of C_p with temperature along several isobars is illustrated in Figure 18, while Figure 19 shows the density dependence of C_v along various isotherms.

124	7.649	32.058	0.23
126	9.288	31.895	0.23
128	10.861	31.727	0.25
130	12.373	31.556	0.25
132	13.824	31.381	0.26
134	15.278	31.214	0.26
136	16.583	31.024	0.28
138	17.980	30.864	0.27
140	19.155	30.664	0.26
142	20.811	30.573	0.31
144	21.668	30.322	0.31

estimated pressure errors are due to a 1 percent
ed uncertainty in either the isotherm or the isochore
tive.

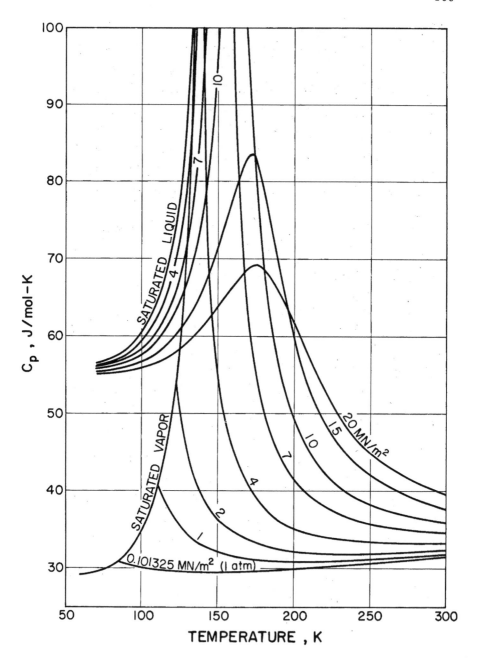

Figure 18. Specific heat at constant pressure, C_p, along selected isobars.

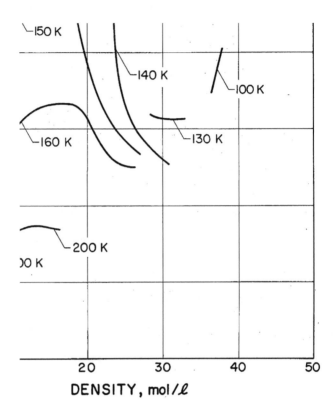

19. Specific heat at constant volume, C_v, along selected isotherms.

E.   Discussions Concerning the Derived Properties.

The best estimates of the accuracy of the derived thermodynamic properties are obtained by making comparisons with experimental values for these properties. In general, though, such
data are often scarce, limited in scope, or of poor accuracy. In
the case of fluorine, thermodynamic property data are practically
nonexistent.

Internal consistency checks may be made by performing
closed loop computations for different regions of the PVT surface.
Because of the somewhat roundabout method used to calculate the
enthalpy of the liquid as described in Section B of this chapter, it
is particularly desirable to make comparisons with liquid enthalpies
obtained by different methods. This may be done by comparing
heats of vaporization reported by different sources as given in
Table 12. The second and third columns of this table present
values of ΔH and $T\Delta S$ obtained from Table 15. The differences
between these two columns need to be explained. ΔH should not
differ from $T\Delta S$ by more than a few J/mol if all correlations used
for the computations provide the "true" value for a particular
variable. An error offset in the vapor pressure at the triple point
of 2 percent would alter the $T\Delta S$ value at this temperature by
9 J/mol due to a change in the entropy value of the vapor as a result of the $R \ln (RT\rho/P_0)$ term in Equation (55). Obviously, ΔH in

T K	ΔH	TΔS	Clapeyron Equation	Clapeyron Equation	Clapeyron Equation
53.481	7578	7543	7506	7510	
60	7407	7377	7341	7323	
70	7130	7106	7060	7032	
80	6825	6807	6759	6714	
84.71	6667	6652	6605		
84.93	6659	6644	6597		
84.95	6659	6644	6597		
85.03	6656	6641	6594	6532	
85.19	6650	6635	6589		
85.21	6649	6635	6588		
90	6475	6462	6418		6116
100	6056	6048	6009		5648
110	5539	5534	5501		5117
120	4877	4875	4852		4472
130	3977	3976	3965		3636
140	2481	2481	2472		2247
144	943	943	935		564

* Calorimetric data.

the second column of Table 12 would not be altered. The differences, however, may also be due to errors in the saturated liquid

specific heat, C_{sat} (Equation (68)). An average error in C_{sat} of

-2 percent between 53.5 K and 135 K would change ΔH at the triple

point temperature by about -103 J/mol and $T\Delta S$ by approximately

-60 J/mol, resulting in a net difference of essentially 43 J/mol.

Other heats of vaporization in Table 12 are calculated by the

Clapeyron equation

$$\Delta H = T(V_g - V_\ell) \frac{dP}{dT} , \qquad (74)$$

where $(V_g - V_\ell)$ is the difference in the molar volumes of the

saturated gaseous and liquid phases and dP/dT is the slope of the

vapor pressure curve. These ΔH - values, however, suffer from

relatively large uncertainties in dP/dT and V_g at low tempera-

tures. Calculations by Ziegler and Mullins [44] agree well with

the new data (ΔH_{clap}) near the triple point but the deviations increas

as the temperature approaches 85 K. This is to be expected since

Ziegler and Mullins' vapor pressures are based on the normal

boiling point of Reference [39], as may be noted from Figure 15.

Fricke [35] based his ΔH - calculations on the Berthelot equation of

state which gives vapor densities differing from Equation (31) by as

much as 10 percent at 140 K.

The entropy value of the saturated liquid at the triple point

calculated using the method described in Section B of this chapter,

that the agreement among the three values is consistent with the Third Law of Thermodynamics. Thus, the $T\Delta S$ values are more likely to represent the correct values of the heats of vaporization since large errors in the vapor pressure equation are not probable.

Experimental specific heats, C_v, measured by Prydz and Goodwin [48] agree well with values calculated from the PVT surface for densities above the critical and temperatures below 160.01 K. This is to be expected since the calculated specific heats in this region of the PVT surface were based on the experimental values for C_v [48] and the C_{sat} measurements of Goodwin and Prydz [57]. Over the rest of the surface the measured C_v values are larger than the values calculated from the PVT data. The disagreement is within the combined uncertainty of the two experiments, although at high pressures (20 MN/m^2) and 160.01 K maximum discrepancies of about 3 percent in C_v are apparent.

Calculated specific heats at constant pressure, C_p, are not as smooth as may be desired. These quantities are based on the ratio of the square of the isochore derivative, $(\partial P/\partial T)_\rho^2$, to the isotherm derivative, $(\partial P/\partial \rho)_T$, which are not likely to be completely smooth in the terminal regions of the PVT surface. In these regions a density error of 0.01 percent can result in relatively large errors in the slopes. Also, the integration of the second derivative of the isochores as indicated by Equation (62) increases the probability of a somewhat erratic behavior. The actual values of the C_p's, however, should be as accurate as may be calculated from an equation of state and indicate better what uncertainty in the computed values may be expected from the calculations.

No experimental sound velocities for fluorine have been reported in the literature. However, the divergence of C_v which has been observed for several fluids, including fluorine [48], near the critical point, implies the vanishing of the isentropic sound velocity as proposed by Yang and Yang [58]. Recent measurements by Barmatz [59] on He^4 also show similar behavior. Calculated velocities of sound for fluorine presented in Tables 15 and 16, tend to indicate such a trend.

XII CONCLUSIONS

An apparatus has been constructed and used successfully to measure about 150 vapor pressure data points and more than 1000 PVT points for fluorine from the triple point to 300 K at pressures to about 24 MN/m². The problems associated with the high reactivity and extreme toxicity of this element which impose severe requirements on system compatibility and provisions for personal safety, were solved.

A network of isotherms and isochores and a virial equation were adopted to represent all PVT data. These equations represent the data to within its precision, which is of the order of 0.05 percent in density over the whole PVT surface. The actual accuracy of the densities predicted from the equations is estimated to be about 0.1 percent. Uncertainties in the temperature and pressure measurements are about 0.01 percent, increasing somewhat in the lower terminal regions. In the critical region the densities obtained at various pressures are necessarily less accurate because of the high isothermal compressibility. However, even though the nonanalytic character that causes the specific heats to diverge in the vicinity of the critical point has not been included in the form of the isotherm and isochore equations, tabulated values of the derived properties given in Tables 15 and 16 are believed to qualitatively show the right behavior in this region. However, caution should be exercised near the critical point when extrapolating or interpolating on the derived properties. This is particularly true for both the heat capacities and the velocity of sound since these properties supposedly exhibit anomalies in this part of the surface.

Due to the fact that this study is relatively complete (it is also based on experimental specific heats), it is believed that measurements from only a few other experiments would help to confirm or improve the accuracy of the PVT data and the derived properties. Some additional measurements utilizing other experimental techniques [60, 61, 62] covering selected portions of the surface would be valuable for the purpose of checking the accuracy of the PVT correlations obtained in this study. PVT measurements covering the critical region would also help to improve the surface. Such data were recently obtained by Weber [63] in his critical region studies of oxygen by means of dielectric measurements. Corresponding measurements for fluorine would be possible since a compatible insulating material for the electrical leads entering the pressure bomb, has now been found [64]. Thus, the critical parameters could be evaluated with a technique that is more accurate than the one used in this study. Finally, very accurate (preferably better than 0.1 percent) experimental heats of vaporization measurements would help to clarify as to which ΔH - column in Table 12 gives the more correct values.

9	0.000301	0.00297
7	0.000410	0.00404
5	0.000534	0.00527
5	0.000540	0.00533
3	0.000712	0.00702
3	0.000713	0.00704
2	0.000928	0.00916
2	0.000928	0.00916
0	0.00120	0.0118
0	0.00120	0.0119
9	0.00155	0.0153
9	0.00154	0.0153
9	0.00155	0.0153
9	0.00197	0.0195
8	0.00249	0.0245
9	0.00311	0.0307
9	0.00387	0.0382
9	0.00386	0.0380
0	0.00477	0.0471
0	0.00478	0.0471
0	0.00585	0.0577
1	0.00710	0.0701
1	0.00859	0.0847
1	0.01030	0.1017
0	0.01231	0.1215
0	0.01231	0.1215
9	0.01485	0.1466

TABLE 13. EXPERIMENTAL VAPOR PRESSURES OF LIQUID
FLUORINE--Continued

T, K	P, MN/m^2	P, atm
71.998	0.01725	0.1702
72.997	0.02026	0.2000
73.996	0.02369	0.2338
74.994	0.02758	0.2722
74.994	0.02760	0.2724
75.993	0.03196	0.3155
76.992	0.03688	0.3640
76.992	0.03690	0.3642
77.991	0.04241	0.4186
78.991	0.04858	0.4794
79.991	0.05540	0.5467
79.991	0.05540	0.5468
80.991	0.06297	0.6215
81.992	0.07135	0.7041
82.994	0.08059	0.7953
83.995	0.09072	0.8953
84.997	0.10186	1.0053
84.997	0.10187	1.0054
86.000	0.11404	1.1255
87.002	0.12730	1.2564
88.005	0.14172	1.3987
88.005	0.14169	1.3983
88.005	0.14170	1.3985
89.007	0.15726	1.5521
89.007	0.15740	1.5534
90.010	0.17420	1.7192
90.010	0.17424	1.7196

EXPERIMENTAL VAPOR PRESSURES OF LIQUID
FLUORINE--Continued

P, MN/m^2	P, atm
0.17420	1.7192
0.17415	1.7187
0.19250	1.8998
0.19240	1.8988
0.21204	2.0927
0.21204	2.0927
0.23320	2.3015
0.25591	2.5256
0.28018	2.7651
0.30610	3.0210
0.33389	3.2952
0.36356	3.5880
0.39497	3.8981
0.42839	4.2279
0.42848	4.2288
0.46390	4.5783
0.50157	4.9501
0.54132	5.3424
0.58345	5.7582
0.62799	6.1978
0.67481	6.6599
0.72434	7.1487
0.77623	7.6608
0.83094	8.2007

TABLE 13. EXPERIMENTAL VAPOR PRESSURES OF LIQUID
FLUORINE--Continued

T, K	P, MN/m^2	P, atm
112.995	1.07805	10.6396
113.995	1.14730	11.3230
114.994	1.21970	12.0375
114.994	1.21971	12.0376
115.993	1.29551	12.7857
116.992	1.37439	13.5641
117.991	1.45689	14.3784
118.990	1.54270	15.2253
119.989	1.63224	16.1090
119.989	1.63234	16.1099
119.989	1.63220	16.1086
120.989	1.72550	17.0293
121.988	1.82242	17.9859
122.987	1.92335	18.9820
123.987	2.02779	20.0128
124.987	2.13705	21.0911
125.987	2.24992	22.2050
126.987	2.36742	23.3646
127.987	2.48879	24.5624
128.987	2.61529	25.8109
129.987	2.74590	27.0999
129.987	2.74548	27.0957
130.987	2.88128	28.4361
131.987	3.02153	29.8202
132.987	3.16649	31.2508
133.988	3.31707	32.7370
134.988	3.47235	34.2694
135.989	3.63326	35.8575

. EXPERIMENTAL VAPOR PRESSURES OF LIQ
FLUORINE--Continued

	P, MN/m^2	P, atm
0	3.80012	37.5043
0	3.97266	39.2071
1	4.15065	40.9637
1	4.15115	40.9687
2	4.33565	42.7896
2	4.33586	42.7916
2	4.33565	42.7895
2	4.52689	44.6769
2	4.52711	44.6791
3	4.72527	46.6348
3	4.72518	46.6339
4	4.93044	48.6597
4	4.93038	48.6590
4	5.03628	49.7043
	*5.215	*51.47

TABLE 14   TEMPERATURE-PRESSURE-DENSITY OBSERVATIONS ON FLUORINE

TEMP K	PRESSURE MN/M²	DENSITY MOL/L	IDENT
74.994	0.0231	0.0379	13101
79.991	0.0249	0.0378	13102
84.997	0.0265	0.0378	13103
90.010	0.0280	0.0378	13104
95.013	0.0296	0.0378	13105
100.010	0.0312	0.0378	13106
109.998	0.0343	0.0377	13107
119.989	0.0374	0.0377	13108
129.987	0.0405	0.0377	13109
139.992	0.0436	0.0376	13110
79.991	0.0479	0.0740	13001
84.997	0.0513	0.0739	13002
90.010	0.0544	0.0739	13003
95.013	0.0575	0.0738	13004
100.010	0.0606	0.0738	13005
109.998	0.0667	0.0737	13006
119.989	0.0729	0.0737	13007
129.987	0.0789	0.0736	13008
139.992	0.0850	0.0735	13009
84.997	0.0955	0.1403	12901
90.010	0.1020	0.1403	12902
95.013	0.1080	0.1402	12903
100.010	0.1139	0.1401	12904
105.004	0.1198	0.1401	12905
119.989	0.1373	0.1398	12906
129.987	0.1489	0.1397	12907
139.992	0.1605	0.1396	12908
90.010	0.1516	0.2118	12801
95.013	0.1609	0.2117	12802
100.010	0.1699	0.2116	12803
105.004	0.1789	0.2115	12804
109.998	0.1879	0.2114	12805
119.989	0.2057	0.2111	12806
129.987	0.2234	0.2109	12807
139.992	0.2410	0.2107	12808
95.013	0.2640	0.3576	12301
100.010	0.2803	0.3574	12302
109.998	0.3112	0.3571	12303
119.989	0.3417	0.3567	12304
129.987	0.3721	0.3563	12305
139.992	0.4022	0.3559	12306
102.008	0.4305	0.5538	12701
105.004	0.4454	0.5537	12702
109.998	0.4699	0.5534	12703
114.994	0.4942	0.5531	12704
119.989	0.5183	0.5528	12705
129.987	0.5661	0.5522	12706
139.992	0.6136	0.5516	12707
105.004	0.6031	0.7767	12501
109.998	0.6390	0.7763	12502
114.994	0.6740	0.7759	12503
119.989	0.7086	0.7755	12504
124.987	0.7430	0.7751	12505
129.987	0.7772	0.7746	12506
139.992	0.8451	0.7738	12507
111.996	0.9141	1.1436	12402
114.994	0.9465	1.1432	12403
119.989	0.9997	1.1426	12404
129.987	1.1044	1.1413	12405
139.992	1.2075	1.1400	12406

TEMP K	PRESSURE MN/M²	DENSITY MOL/L
115.993	1.2050	1.5135
117.991	1.2348	1.5132
119.989	1.2641	1.5128
121.988	1.2932	1.5125
123.987	1.3221	1.5121
125.987	1.3510	1.5118
127.987	1.3794	1.5114
129.987	1.4080	1.5111
131.987	1.4363	1.5107
133.988	1.4645	1.5104
135.989	1.4925	1.5101
137.990	1.5206	1.5097
139.992	1.5486	1.5094
141.993	1.5765	1.5090
143.995	1.6042	1.5087
145.996	1.6319	1.5083
147.998	1.6596	1.5080
150.000	1.6871	1.5077
155.005	1.7556	1.5068
160.010	1.8237	1.5059
165.014	1.8916	1.5051
170.019	1.9592	1.5042
175.024	2.0264	1.5033
180.027	2.0936	1.5025
185.030	2.1603	1.5016
190.032	2.2268	1.5007
195.034	2.2931	1.4998
200.035	2.3592	1.4990
210.033	2.4909	1.4972
220.030	2.6218	1.4954
220.030	2.6214	1.4954
230.025	2.7517	1.4935
240.020	2.8812	1.4917
250.014	3.0103	1.4898
260.008	3.1389	1.4880
270.002	3.2670	1.4861
279.997	3.3946	1.4842
289.994	3.5216	1.4822
299.992	3.6484	1.4803
121.988	1.6805	2.1162
124.987	1.7449	2.1154
129.987	1.8504	2.1141
134.988	1.9543	2.1129
139.992	2.0568	2.1116
144.995	2.1583	2.1104
150.000	2.2589	2.1091
125.987	2.0758	2.6479
127.987	2.1314	2.6472
129.987	2.1865	2.6466
131.987	2.2411	2.6459
133.988	2.2954	2.6453
135.989	2.3491	2.6446
137.990	2.4026	2.6440
139.992	2.4558	2.6433
141.993	2.5088	2.6427
143.995	2.5614	2.6420
145.996	2.6138	2.6414
147.998	2.6659	2.6407
150.000	2.7180	2.6401
155.005	2.8470	2.6384
160.010	2.9751	2.6368
165.014	3.1021	2.6352
170.019	3.2284	2.6336
175.024	3.3538	2.6319
180.027	3.4784	2.6303
185.030	3.6024	2.6287
190.032	3.7260	2.6270
195.034	3.8486	2.6254

TABLE 14 TEMPERATURE-PRESSURE-DENSITY OBSERVATIONS ON FLUORINE..CONTI

TEMP K	PRESSURE MN/M²	DENSITY MOL/L	IDENT
195.034	3.8486	2.6254	7422
200.035	3.9711	2.6238	7423
210.033	4.2145	2.6204	7424
220.030	4.4562	2.6171	7425
230.025	4.6964	2.6138	7426
240.020	4.9347	2.6104	7427
230.025	4.6968	2.6140	7428
240.020	4.9356	2.6106	7429
250.014	5.1732	2.6072	7430
260.008	5.4093	2.6038	7431
270.002	5.6446	2.6004	7432
279.997	5.8785	2.5969	7433
289.994	6.1114	2.5933	7434
299.992	6.3432	2.5897	7435
129.987	2.6958	3.6637	7201
131.987	2.7788	3.6627	7202
133.988	2.8604	3.6617	7203
135.989	2.9410	3.6608	7204
137.990	3.0209	3.6598	7205
139.992	3.1001	3.6588	7206
141.993	3.1786	3.6579	7207
143.995	3.2565	3.6569	7208
145.996	3.3341	3.6560	7209
147.998	3.4110	3.6550	7210
150.000	3.4876	3.6540	7211
155.005	3.6773	3.6517	7212
160.010	3.8649	3.6493	7213
165.014	4.0507	3.6470	7214
170.019	4.2350	3.6446	7215
175.024	4.4179	3.6423	7216
180.027	4.5994	3.6399	7217
180.027	4.5995	3.6401	7218
185.030	4.7802	3.6378	7219
190.032	4.9597	3.6354	7220
195.034	5.1383	3.6330	7221
200.035	5.3160	3.6307	7222
210.033	5.6689	3.6259	7223
220.030	6.0188	3.6211	7224
230.025	6.3662	3.6163	7225
240.020	6.7110	3.6114	7226
250.014	7.0537	3.6065	7227
260.008	7.3946	3.6015	7228
270.002	7.7335	3.5966	7229
279.997	8.0706	3.5915	7230
289.994	8.4063	3.5865	7231
299.992	8.7404	3.5813	7232
150.000	3.9137	4.3084	8501
165.014	4.6013	4.2998	8502
180.027	5.2692	4.2913	8503
200.035	6.1393	4.2799	8504
230.025	7.4114	4.2626	8505
133.988	3.2925	4.7443	7001
135.989	3.4060	4.7430	7002
137.990	3.5176	4.7416	7003
139.992	3.6277	4.7403	7004
141.993	3.7366	4.7389	7005
143.995	3.8445	4.7376	7006
145.996	3.9514	4.7363	7007
147.998	4.0575	4.7350	7008
150.000	4.1629	4.7337	7009
155.005	4.4233	4.7304	7010
160.010	4.6804	4.7272	7011

TEMP K	PRESSURE MN/M²	DENSITY MOL/L
165.014	4.9344	4.7240
170.019	5.1861	4.7208
175.024	5.4345	4.7176
180.027	5.6819	4.7144
185.030	5.9278	4.7113
190.032	6.1719	4.7081
195.034	6.4144	4.7049
200.035	6.6558	4.7018
210.033	7.1347	4.6954
220.030	7.6091	4.6890
210.033	7.1344	4.6952
220.030	7.6091	4.6888
230.025	8.0796	4.6824
240.020	8.5464	4.6759
250.014	9.0103	4.6693
260.008	9.4713	4.6627
270.002	9.9295	4.6560
279.997	10.3848	4.6493
289.994	10.8367	4.6425
299.992	11.2872	4.6357
137.990	3.7717	5.4871
139.992	3.9057	5.4855
141.993	4.0376	5.4839
143.995	4.1679	5.4823
145.996	4.2969	5.4808
147.998	4.4249	5.4792
150.000	4.5515	5.4776
155.005	4.8645	5.4737
160.010	5.1727	5.4699
165.014	5.4770	5.4660
170.019	5.7784	5.4622
175.024	6.0765	5.4585
180.027	6.3724	5.4547
185.030	6.6661	5.4509
190.032	6.9578	5.4471
195.034	7.2477	5.4433
200.035	7.5357	5.4395
210.033	8.1071	5.4319
220.030	8.6730	5.4242
220.030	8.6718	5.4241
230.025	9.2328	5.4165
240.020	9.7891	5.4087
250.014	10.3417	5.4008
260.008	10.8902	5.3929
270.002	11.4359	5.3849
279.997	11.9776	5.3769
289.994	12.5164	5.3687
299.992	13.0526	5.3605
137.990	3.9283	6.0824
139.992	4.0827	6.0805
141.993	4.2341	6.0787
143.995	4.3832	6.0769
145.996	4.5308	6.0751
147.998	4.6767	6.0733
150.000	4.8213	6.0715
155.005	5.1779	6.0671
160.010	5.5288	6.0627
165.014	5.8752	6.0583
170.019	6.2174	6.0540
175.024	6.5565	6.0496
180.027	6.8924	6.0453
185.030	7.2260	6.0410
190.032	7.5572	6.0366
195.034	7.8860	6.0323

TABLE 14   TEMPERATURE-PRESSURE-DENSITY OBSERVATIONS ON FLUORINE.-CONTINUED

TEMP K	PRESSURE MN/M²	DENSITY MOL/L	IDENT	TEMP K	PRESSURE MN/M²	DENSITY MOL/L	IDENT
200.035	8.2131	6.0279	6917	289.994	15.6802	6.7219	2727
210.033	8.8617	6.0192	6918	299.992	16.3876	6.7111	2728
220.030	9.5038	6.0105	6919				
230.025	10.1405	6.0017	6920				
240.020	10.7703	5.9929	6921	141.993	4.5958	7.6840	7501
250.014	11.3966	5.9841	6922	143.995	4.8006	7.6815	7502
260.008	12.0190	5.9751	6923	145.996	5.0015	7.6791	7503
270.002	12.6373	5.9662	6924	147.998	5.2001	7.6767	7504
279.997	13.2518	5.9571	6925	150.000	5.3960	7.6743	7505
289.994	13.8633	5.9479	6926	155.005	5.8780	7.6683	7506
299.992	14.4706	5.9387	6927	160.010	6.3513	7.6624	7507
				165.014	6.8179	7.6565	7508
				170.019	7.2790	7.6507	7509
139.992	4.2052	6.5928	7801	175.024	7.7353	7.6449	7510
141.993	4.3745	6.5908	7802	180.027	8.1862	7.6392	7511
143.995	4.5407	6.5888	7803	185.030	8.6352	7.6334	7512
145.996	4.7050	6.5868	7804	190.032	9.0807	7.6276	7513
147.998	4.8670	6.5848	7805	190.032	9.0802	7.6275	7514
150.000	5.0275	6.5828	7806	195.034	9.5228	7.6217	7515
155.005	5.4232	6.5779	7807	200.035	9.9626	7.6159	7516
160.010	5.8120	6.5730	7808	210.033	10.8353	7.6043	7517
165.014	6.1955	6.5681	7809	220.030	11.6987	7.5926	7518
170.019	6.5732	6.5633	7810	230.025	12.5549	7.5808	7519
175.024	6.9484	6.5586	7811	240.020	13.4022	7.5689	7520
180.027	7.3206	6.5538	7812	250.014	14.2444	7.5570	7521
185.030	7.6898	6.5490	7813	260.008	15.0797	7.5449	7522
190.032	8.0560	6.5442	7814	270.002	15.9091	7.5327	7523
195.034	8.4200	6.5395	7815	279.997	16.7342	7.5203	7524
200.035	8.7819	6.5346	7816	289.994	17.5523	7.5078	7525
210.033	9.4992	6.5251	7817	299.992	18.3667	7.4953	7526
220.030	10.2098	6.5154	7818				
230.025	10.9136	6.5058	7819				
230.025	10.9148	6.5061	7820	185.030	8.9694	8.0302	8401
240.020	11.6123	6.4962	7821	195.034	9.9172	8.0179	8402
250.014	12.3046	6.4862	7822	210.033	11.3172	7.9994	8403
260.008	12.9920	6.4762	7823	230.025	13.1530	7.9743	8404
270.002	13.6750	6.4661	7824				
279.997	14.3536	6.4559	7825				
289.994	15.0284	6.4456	7826	143.995	4.9168	8.3819	11901
299.992	15.6989	6.4353	7827	145.996	5.1430	8.3791	11902
				147.998	5.3652	8.3764	11903
139.992	4.2620	6.8747	2701	150.000	5.5847	8.3737	11904
141.993	4.4415	6.8725	2702	155.005	6.1242	8.3671	11905
143.995	4.6175	6.8704	2703	160.010	6.6540	8.3605	11906
145.996	4.7909	6.8683	2704	165.014	7.1759	8.3539	11907
147.998	4.9623	6.8662	2705	170.019	7.6922	8.3474	11908
150.000	5.1318	6.8642	2706	175.024	8.2027	8.3409	11909
155.005	5.5490	6.8590	2707	180.027	8.7093	8.3345	11910
160.010	5.9593	6.8540	2708	185.030	9.2117	8.3280	11911
165.014	6.3637	6.8489	2709	190.032	9.7107	8.3215	11912
170.019	6.7637	6.8439	2710	195.034	10.2065	8.3150	11913
175.024	7.1593	6.8389	2711	200.035	10.6993	8.3085	11914
180.027	7.5515	6.8339	2712	210.033	11.6772	8.2955	11915
185.030	7.9406	6.8290	2713	220.030	12.6404	8.2825	11916
190.032	8.3270	6.8240	2714	210.033	11.6741	8.2952	11917
195.034	8.7109	6.8190	2715	220.030	12.6416	8.2821	11918
200.035	9.0924	6.8140	2716	230.025	13.5995	8.2689	11919
210.033	9.8489	6.8040	2717	240.020	14.5502	8.2556	11920
220.030	10.5977	6.7940	2718	250.014	15.4923	8.2422	11921
230.025	11.3398	6.7839	2719	260.008	16.4281	8.2287	11922
220.030	10.5990	6.7943	2720	270.002	17.3573	8.2151	11923
230.025	11.3409	6.7842	2721	279.997	18.2803	8.2013	11924
240.020	12.0771	6.7740	2722	289.994	19.1973	8.1874	11925
250.014	12.8074	6.7638	2723	299.992	20.1092	8.1732	11926
260.008	13.5329	6.7535	2724				
270.002	14.2532	6.7430	2725				
279.997	14.9689	6.7325	2726	141.993	4.7140	8.7499	3102

TABLE 14   TEMPERATURE-PRESSURE-DENSITY OBSERVATIONS ON FLUORINE--CONTINUED

TEMP K	PRESSURE MN/M^2	DENSITY MOL/L	IDENT	TEMP K	PRESSURE MN/M^2	DENSITY MOL/L	IDENT
43.995	4.9677	8.7468	3103	200.035	12.1356	9.6809	7614
45.996	5.2069	8.7439	3104	210.033	13.3343	9.6649	7615
47.998	5.4421	8.7411	3105	220.030	14.5208	9.6488	7616
50.000	5.6739	8.7382	3106	230.025	15.6937	9.6326	7617
55.005	6.2440	8.7312	3107	230.025	15.6950	9.6329	7618
60.010	6.8033	8.7243	3108	240.020	16.8613	9.6166	7619
65.014	7.3548	8.7174	3109	250.014	18.0166	9.6000	7620
70.019	7.9002	8.7105	3110	260.008	19.1645	9.5833	7621
75.024	8.4399	8.7037	3111	270.002	20.3049	9.5663	7622
80.027	8.9755	8.6969	3112				
85.030	9.5067	8.6901	3113				
90.032	10.0342	8.6833	3114	143.995	5.1231	10.8686	8001
95.034	10.5584	8.6765	3115	145.996	5.4415	10.8647	8002
00.035	11.0798	8.6697	3116	147.998	5.7518	10.8608	8003
10.033	12.1133	8.6560	3117	150.000	6.0578	10.8570	8004
20.030	13.1363	8.6423	3118	155.005	6.8106	10.8477	8005
20.030	13.1341	8.6416	3119	160.010	7.5519	10.8384	8006
30.025	14.1478	8.6278	3120	165.014	8.2847	10.8291	8007
40.020	15.1530	8.6140	3121	170.019	9.0106	10.8199	8008
50.014	16.1502	8.6000	3122	175.024	9.7306	10.8108	8009
60.008	17.1403	8.5859	3123	180.027	10.4459	10.8016	8010
70.002	18.1231	8.5717	3124	185.030	11.1567	10.7925	8011
79.997	19.0990	8.5572	3125	190.032	11.8630	10.7834	8012
89.994	20.0680	8.5426	3126	195.034	12.5659	10.7742	8013
99.992	21.0320	8.5278	3127	200.035	13.2650	10.7650	8014
				210.033	14.6522	10.7466	8015
				220.030	16.0262	10.7279	8016
43.995	4.9995	9.0187	8201	230.025	17.3881	10.7092	8017
45.996	5.2489	9.0157	8202	240.020	18.7385	10.6903	8018
47.998	5.4935	9.0127	8203	250.014	20.0784	10.6712	8019
50.000	5.7349	9.0097	8204	260.008	21.4083	10.6519	8020
55.005	6.3276	9.0024	8205				
60.010	6.9099	8.9953	8206				
65.014	7.4838	8.9881	8207	145.996	5.5018	11.9244	7902
70.019	8.0512	8.9809	8208	147.998	5.8491	11.9201	7903
75.024	8.6132	8.9738	8209	150.000	6.1922	11.9159	7904
80.027	9.1705	8.9667	8210	155.005	7.0392	11.9053	7905
85.030	9.7235	8.9596	8211	160.010	7.8758	11.8948	7906
90.032	10.2729	8.9524	8212	165.014	8.7053	11.8843	7907
95.034	10.8190	8.9453	8213	170.019	9.5289	11.8739	7908
00.035	11.3619	8.9381	8214	175.024	10.3471	11.8634	7909
43.995	4.9977	9.0186	8215	180.027	11.1599	11.8530	7910
70.019	8.0504	8.9805	8216	185.030	11.9686	11.8425	7911
00.035	11.3610	8.9378	8217	190.032	12.7733	11.8321	7912
10.033	12.4354	8.9234	8218	195.034	13.5742	11.8216	7913
20.030	13.5009	8.9090	8219	200.035	14.3705	11.8111	7914
30.025	14.5573	8.8945	8220	210.033	15.9527	11.7900	7915
40.020	15.6029	8.8798	8221	220.030	17.5202	11.7687	7916
50.014	16.6645	8.8650	8222	230.025	19.0742	11.7472	7917
60.008	17.6717	8.8501	8223	240.020	20.6152	11.7254	7918
70.002	18.6956	8.8349	8224				
79.997	19.7114	8.8197	8225				
89.994	20.7215	8.8043	8226	145.996	5.5529	13.4190	7701
				147.998	5.9507	13.4141	7702
				150.000	6.3460	13.4092	7703
43.995	5.0676	9.7709	7601	155.005	7.3299	13.3969	7704
45.996	5.3447	9.7675	7602	160.010	8.3087	13.3846	7705
47.998	5.6161	9.7641	7603	165.014	9.2833	13.3724	7706
50.000	5.8835	9.7608	7604	170.019	10.2538	13.3601	7707
55.005	6.5404	9.7527	7605	175.024	11.2199	13.3478	7708
60.010	7.1858	9.7446	7606	180.027	12.1821	13.3355	7709
65.014	7.8231	9.7366	7607	185.030	13.1408	13.3232	7710
70.019	8.4535	9.7286	7608	190.032	14.0952	13.3108	7711
75.024	9.0778	9.7206	7609	195.034	15.0452	13.2983	7712
80.027	9.6972	9.7127	7610	200.035	15.9911	13.2858	7713
85.030	10.3125	9.7047	7611	210.033	17.8715	13.2606	7714

TABLE 14   TEMPERATURE-PRESSURE-DENSITY OBSERVATIONS ON FLUORINE.--CONTINUED

TEMP K	PRESSURE MN/M²	DENSITY MOL/L	IDENT	TEMP K	PRESSURE MN/M²	DENSITY MOL/L	IDENT
145.996	5.5803	14.4938	7301	185.030	19.8847	19.6335	9108
147.998	6.0136	14.4884	7302	190.032	21.6917	19.6097	9109
150.000	6.4479	14.4829	7303				
155.005	7.5346	14.4693	7304				
160.010	8.6223	14.4556	7305	143.995	5.2963	20.6673	11401
165.014	9.7090	14.4418	7306	145.996	6.0273	20.6582	11402
170.019	10.7932	14.4280	7307	147.998	6.7749	20.6488	11403
175.024	11.8746	14.4143	7308	150.000	7.5327	20.6392	11404
180.027	12.9527	14.4004	7309	155.005	9.4574	20.6149	11405
185.030	14.0275	14.3865	7310	160.010	11.4063	20.5904	11406
190.032	15.0986	14.3726	7311	165.014	13.3667	20.5659	11407
195.034	16.1651	14.3586	7312	170.019	15.3332	20.5411	11408
200.035	17.2279	14.3445	7313	175.024	17.2983	20.5162	11409
210.033	19.3406	14.3161	7314	180.027	19.2617	20.4911	11410
220.030	21.4361	14.2873	7315	185.030	21.2213	20.4656	11411
145.996	5.6063	15.7701	7102	143.995	5.5541	21.8571	11701
147.998	6.0859	15.7641	7103	145.996	6.3880	21.8467	11702
150.000	6.5689	15.7580	7104	147.998	7.2345	21.8361	11703
155.005	7.7889	15.7427	7105	150.000	8.0903	21.8253	11704
160.010	9.0163	15.7272	7106	155.005	10.2564	21.7981	11705
165.014	10.2477	15.7118	7107	160.010	12.4430	21.7707	11706
170.019	11.4795	15.6962	7108	165.014	14.6396	21.7433	11707
175.024	12.7098	15.6806	7109	170.019	16.8381	21.7157	11708
180.027	13.9387	15.6650	7110	175.024	19.0347	21.6877	11709
185.030	15.1641	15.6493	7111	180.027	21.2276	21.6595	11710
190.032	16.3858	15.6335	7112				
195.034	17.6042	15.6175	7113				
200.035	18.8182	15.6015	7114	143.995	5.5550	21.8658	11301
210.033	21.2328	15.5691	7115	145.996	6.3896	21.8554	11302
				147.998	7.2372	21.8448	11303
				150.000	8.0928	21.8339	11304
145.996	5.6826	17.9228	11801	155.005	10.2602	21.8066	11305
147.998	6.2544	17.9156	11802	160.010	12.4475	21.7792	11306
150.000	6.8366	17.9083	11803	165.014	14.6451	21.7516	11307
155.005	8.3181	17.8896	11804	170.019	16.8455	21.7238	11308
160.010	9.8205	17.8706	11805	175.024	19.0414	21.6959	11309
165.014	11.3329	17.8516	11806	180.027	21.2361	21.6675	11310
170.019	12.8497	17.8324	11807				
175.024	14.3668	17.8131	11808				
180.027	15.8834	17.7937	11809	141.993	5.0864	23.0236	6801
185.030	17.3973	17.7742	11810	143.995	6.0185	23.0124	6802
190.032	18.9083	17.7545	11811	145.996	6.9648	23.0006	6803
195.034	20.4147	17.7346	11812	147.998	7.9207	22.9885	6804
				150.000	8.8832	22.9762	6805
				155.005	11.3096	22.9458	6806
145.996	5.7462	18.7458	11501	160.010	13.7511	22.9154	6807
147.998	6.3635	18.7380	11502	165.014	16.1970	22.8849	6808
150.000	6.9921	18.7301	11503	170.019	18.6432	22.8541	6809
155.005	8.5924	18.7099	11504	175.024	21.0844	22.8229	6810
160.010	10.2157	18.6895	11505				
165.014	11.8507	18.6689	11506				
170.019	13.4911	18.6482	11507	139.992	4.6408	24.1656	11201
175.024	15.1326	18.6273	11508	141.993	5.6875	24.1528	11202
180.027	16.7728	18.6063	11509	143.995	6.7473	24.1395	11203
185.030	18.4107	18.5851	11510	145.996	7.8162	24.1259	11204
190.032	20.0447	18.5637	11511	147.998	8.8917	24.1123	11205
195.034	21.6738	18.5422	11512	150.000	9.9710	24.0989	11206
				155.005	12.6845	24.0653	11207
				160.010	15.4054	24.0316	11208
150.000	7.2527	19.7940	9101	165.014	18.1249	23.9977	11209
155.005	9.0246	19.7716	9102	170.019	20.8398	23.9633	11210
160.010	10.8216	19.7489	9103				
165.014	12.6306	19.7262	9104				
170.019	14.4455	19.7032	9105	137.990	4.3981	25.4197	11101
175.024	16.2603	19.6802	9106	139.992	5.5916	25.4051	11102
180.027	18.0749	19.6569	9107	141.993	6.7956	25.3899	11103

TABLE 14   TEMPERATURE-PRESSURE-DENSITY OBSERVATIONS ON FLUORINE--CONTINUE

TEMP K	PRESSURE MN/M²	DENSITY MOL/L	IDENT	TEMP K	PRESSURE MN/M²	DENSITY MOL/L
143.995	8.0063	25.3744	·11104	137.990	15.9751	30.4277
145.996	9.2228	25.3591	11105	139.992	17.9331	30.4041
147.998	10.4416	25.3440	11106	141.993	19.8854	30.3804
150.000	11.6629	25.3290	11107	143.995	21.8300	· 30.3565
155.005	14.7193	25.2915	11108			
160.010	17.7735	25.2537	11109			
165.014	20.8210	25.2154	11110	123.987	2.5342	30.7537
				125.987	4.5717	30.7266
				127.987	6.6241	30.7009
135.989	4.2576	26.5253	6701	129.987	8.6274	30.6770
137.990	5.5957	26.5094	6702	131.987	10.6449	30.6528
139.992	6.9422	26.4921	6703	133.988	12.6542	30.6288
141.993	8.2923	26.4750	6704	135.989	14.6559	30.6048
143.995	9.6452	26.4584	6705	137.990	16.6506	30.5808
145.996	11.0021	26.4420	6706	139.992	18.6381	30.5568
147.998	12.3580	26.4255	6707	141.993	20.6167	30.5326
150.000	13.7136	26.4091	6708			
155.005	17.1002	26.3679	6709			
160.010	20.4746	26.3261	6710	121.988	2.1204	31.2838
				123.987	4.2621	31.2549
				125.987	6.3997	31.2282
135.989	6.1799	27.7894	8601	127.987	8.5275	31.2031
139.992	9.2435	27.7508	8602	129.987	10.6450	31.1779
145.996	13.8358	27.6951	8603	131.987	12.7562	31.1528
150.000	16.8861	27.6579	8604	133.988	14.8557	31.1280
				135.989	16.9441	31.1032
				137.990	19.0306	31.0783
131.987	3.1337	27.8278	3001	139.992	21.1040	31.0533
133.988	4.6569	27.8076	3002			
135.989	6.1838	27.7893	3003			
137.990	7.7142	27.7695	3004	119.989	2.0538	31.8837
139.992	9.2456	27.7509	3005	121.988	4.3243	31.8531
141.993	10.7772	27.7324	3006	123.987	6.5827	31.8252
143.995	12.3076	27.7139	3007	125.987	8.8306	31.7984
145.996	13.8350	27.6954	3008	127.987	11.0664	31.7717
147.998	15.3607	27.6769	3009	129.987	13.2909	31.7452
150.000	16.8833	27.6584	3010	131.987	15.5074	31.7189
55.005	20.6776	27.6116	3011	133.988	17.7116	31.6926
				135.989	19.9058	31.6661
				137.990	22.0927	31.6394
129.987	3.1750	28.7586	6601			
131.987	4.8498	28.7366	6602			
133.988	6.5264	28.7154	6603	119.989	2.0907	31.8967
135.989	8.2033	28.6953	6604	121.988	4.3610	31.8659
37.990	9.8768	28.6752	6605	123.987	6.6209	31.8379
39.992	11.5492	28.6552	6606	125.987	8.8700	31.8109
41.993	13.2188	28.6352	6607	127.987	11.1072	31.7840
43.995	14.8842	28.6153	6608	129.987	13.3333	31.7573
45.996	16.5472	28.5953	6609	131.987	15.5502	31.7308
47.998	18.2059	28.5752	6610	133.988	17.7548	31.7043
50.000	19.8604	28.5551	6611	135.989	19.9523	31.6777
52.002	21.5104	28.5348	6612	137.990	22.1410	31.6508
129.987	5.1513	29.5967	8701	117.991	1.8073	32.4138
133.988	8.7944	29.5529	8702	119.989	4.1910	32.3815
139.992	14.2282	29.4876	8703	121.988	6.5648	32.3525
147.998	21.4010	29.4005	8704	123.987	8.9247	32.3246
				125.987	11.2704	32.2967
				127.987	13.6024	32.2691
123.987	2.0663	30.5991	3301	129.987	15.9198	32.2416
125.987	4.0717	30.5717	3302	131.987	18.2273	32.2141
127.987	6.0736	30.5458	3303	133.988	20.5256	32.1863
129.987	8.0687	30.5223	3304			
131.987	10.0571	30.4985	3305			
133.988	12.0370	30.4748	3306	117.991	3.8594	32.8504
135.989	14.0095	30.4512	3307	121.988	8.8046	32.7903

TABLE 14   TEMPERATURE-PRESSURE-DENSITY OBSERVATIONS ON FLUORINE--CONTINUED

TEMP K	PRESSURE MN/M²	DENSITY MOL/L	IDENT	TEMP K	PRESSURE MN/M²	DENSITY MOL/L	IDENT
125.987	13.6829	32.7320	8803	109.998	4.9055	34.9504	9702
131.987	20.9035	32.6457	8804	111.996	7.8958	34.9153	9703
				113.995	10.8562	34.8796	9704
				115.993	13.7914	34.8451	9705
115.993	1.3862	32.8854	3501	117.991	16.7036	34.8109	9706
117.991	3.8764	32.8507	3502	119.989	19.5986	34.7764	9707
119.989	6.3523	32.8203	3503	121.988	22.4815	34.7415	9708
121.988	8.8139	32.7911	3504				
123.987	11.2577	32.7619	3505				
125.987	13.6896	32.7332	3506	106.002	1.3049	35.3710	5301
127.987	16.1053	32.7047	3507	108.000	4.4265	35.3281	5302
129.987	18.5101	32.6761	3508	109.998	7.5173	35.2916	5303
131.987	20.9029	32.6473	3509	111.996	10.5819	35.2546	5304
				113.995	13.6157	35.2192	5305
				115.993	16.6268	35.1841	5306
113.995	1.4096	33.4281	2801	117.991	19.6241	35.1489	5307
115.993	4.0243	33.3920	2802	119.989	22.6020	35.1132	5308
117.991	6.6219	33.3603	2803				
119.989	9.2038	33.3296	2804				
121.988	11.7677	33.2990	2805	108.000	4.4294	35.3292	8901
123.987	14.3135	33.2691	2806	111.996	10.5904	35.2558	8902
125.987	16.8497	33.2396	2807	115.993	16.6385	35.1848	8903
127.987	19.3713	33.2097	2808	119.989	22.6077	35.1134	8904
129.987	21.8803	33.1795	2809				
113.995	1.3999	33.4306	9301	104.005	1.7569	35.8759	9801
115.993	4.0187	33.3942	9302	106.002	5.0226	35.8326	9802
117.991	6.6221	33.3625	9303	108.000	8.2535	35.7945	9803
119.989	9.2074	33.3315	9304	109.998	11.4495	35.7565	9804
121.988	11.7742	33.3008	9305	111.996	14.6184	35.7196	9805
123.987	14.3248	33.2707	9306	113.995	17.7614	35.6826	9806
125.987	16.8605	33.2407	9307	115.993	20.8837	35.6453	9807
127.987	19.3844	33.2106	9308				
129.987	21.8939	33.1800	9309				
				102.008	0.9747	36.2163	5501
				104.005	4.3398	36.1697	5502
111.996	1.2522	33.9215	6401	106.002	7.6717	36.1303	5503
113.995	3.9918	33.8833	6402	108.000	10.9641	36.0910	5504
115.993	6.7161	33.8504	6403	109.998	14.2237	36.0531	5505
117.991	9.4152	33.8181	6404	111.996	17.4597	36.0153	5506
119.989	12.0947	33.7862	6405	113.995	20.6766	35.9771	5507
121.988	14.7578	33.7550	6406				
123.987	17.4042	33.7239	6407				
125.987	20.0399	33.6925	6408	102.008	1.2377	36.2449	5701
127.987	22.6598	33.6609	6409	104.005	4.6118	36.1990	5702
				106.002	7.9501	36.1594	5703
				108.000	11.2504	36.1203	5704
111.996	3.7877	34.3287	9401	109.998	14.5134	36.0819	5705
113.995	6.6219	34.2945	9402	111.996	17.7568	36.0440	5706
115.993	9.4336	34.2606	9403	113.995	20.9811	36.0056	5707
117.991	12.2178	34.2276	9404				
119.989	14.9846	34.1951	9405				
121.988	17.7393	34.1625	9406	102.008	3.9537	36.5444	9501
123.987	20.4747	34.1298	9407	104.005	7.4001	36.5033	9502
				106.002	10.7999	36.4627	9503
				108.000	14.1678	36.4237	9504
109.998	0.9643	34.3721	3201	109.998	17.5067	36.3846	950
111.996	3.8173	34.3315	3202	111.996	20.8277	36.3452	950
113.995	6.6497	34.2974	3203				
115.993	9.4560	34.2637	3204				
117.991	12.2408	34.2307	3205	100.010	0.4947	36.5951	290
119.989	15.0052	34.1983	3206	102.008	3.9704	36.5447	290
121.988	17.7529	34.1658	3207	104.005	7.4128	36.5043	290
123.987	20.4844	34.1332	3208	106.002	10.8108	36.4638	290
				108.000	14.1772	36.4249	290
				109.998	17.5172	36.3861	290
108.000	1.8893	34.9904	9701	111.996	20.8393	36.3469	290

TABLE 14   TEMPERATURE-PRESSURE-DENSITY OBSERVATIONS ON FLUORINE--CONTINUED

TEMP K	PRESSURE MN/M²	DENSITY MOL/L	IDENT	TEMP K	PRESSURE MN/M²	DENSITY MOL/L	IDENT
.012	0.8850	37.0590	5101	88.005	4.7505	39.3046	10101
.010	4.5114	37.0089	5102	90.010	9.1162	39.2535	10102
.008	8.0937	36.9662	5103	92.012	13.4113	39.2043	10103
.005	11.6289	36.9252	5104	94.013	17.6633	39.1551	10104
.002	15.1285	36.8847	5105	96.013	21.8822	39.1052	10105
.000	18.6042	36.8442	5106				
.998	22.0614	36.8031	5107				
				83.995	0.4459	39.7405	10501
				86.000	5.0134	39.6764	10502
.012	4.9668	37.4461	9201	88.005	9.5029	39.6242	10503
.008	12.3521	37.3597	9202	90.010	13.9204	39.5735	10504
.005	15.9787	37.3177	9203	92.012	18.3051	39.5229	10505
.002	19.5824	37.2755	9204	94.013	22.6595	39.4713	10506
.000	23.1667	37.2326	9205				
				83.995	0.4651	39.7423	4501
.013	1.3101	37.9022	10001	86.000	5.0379	39.6788	4502
.013	5.2104	37.8533	10002	88.005	9.5411	39.6268	4503
.012	9.0596	37.8054	10003	90.010	13.9741	39.5764	4504
.010	12.8559	37.7618	10004	92.012	18.3629	39.5259	4505
.008	16.6183	37.7182	10005	94.013	22.7325	39.4744	4506
.005	20.3521	37.6744	10006				
				81.992	0.4370	40.1067	10801
.013	1.3182	37.9060	5001	83.995	5.1394	40.0403	10802
.013	5.2200	37.8572	5002	86.000	9.7620	39.9880	10803
.012	9.0684	37.8094	5003	88.005	14.3199	39.9359	10804
.010	12.8621	37.7659	5004	90.010	18.8417	39.8836	10805
.008	16.6237	37.7225	5005	92.012	23.3386	39.8303	10806
.005	20.3536	37.6789	5006				
				81.992	0.4377	40.1072	4201
.012	1.6868	38.3248	9601	83.995	5.1391	40.0409	4202
.013	5.7328	38.2752	9602	86.000	9.7704	39.9887	4203
.013	9.7164	38.2263	9603	88.005	14.3384	39.9367	4204
.012	13.6445	38.1812	9604	90.010	18.8670	39.8846	4205
.010	17.5364	38.1360	9605	92.012	23.3699	39.8314	4206
.008	21.4026	38.0904	9606				
				79.991	0.4890	40.4642	10201
.010	0.9805	38.6564	5201	81.992	5.3375	40.3965	10202
.012	5.1494	38.6036	5202	83.995	10.0991	40.3430	10203
.013	9.2524	38.5525	5203	86.000	14.7862	40.2894	10204
.013	13.2953	38.5062	5204	88.005	19.4396	40.2353	10205
.012	17.3028	38.4601	5205				
.010	21.2828	38.4135	5206				
				79.991	0.4945	40.4640	4401
				81.992	5.3375	40.3965	4402
.010	4.7415	38.9389	9001	83.995	10.0956	40.3432	4403
.013	13.1362	38.8416	9002	86.000	14.7864	40.2897	4404
.012	21.3414	38.7459	9003	88.005	19.4439	40.2359	4405
.005	0.4645	39.0005	4601	77.991	1.4832	40.8761	10301
.010	4.7619	38.9402	4602	79.991	6.4748	40.8127	10302
.012	8.9965	38.8907	4603	81.992	11.3732	40.7571	10303
.013	13.1625	38.8431	4604	83.995	16.2170	40.7016	10304
.013	17.2837	38.7957	4605	86.000	21.0321	40.6455	10305
.012	21.3782	38.7476	4606				
				75.993	1.6279	41.2328	6301
.000	0.3367	39.3667	3401	77.991	6.7642	41.1688	6302
.005	4.7559	39.3037	3402	79.991	11.7882	41.1121	6303
.010	9.1085	39.2529	3403	81.992	16.7735	41.0553	6304
.012	13.3967	39.2040	3404	83.995	21.7365	40.9977	6305
.013	17.6480	39.1552	3405				
.013	21.8622	39.1059	3406				

TABLE 14   TEMPERATURE-PRESSURE-DENSITY OBSERVATIONS ON FLUORINE--CONTINUED

TEMP K	PRESSURE MN/M²	DENSITY MOL/L	IDENT	TEMP K	PRESSURE MN/M²	DENSITY MOL/L	IDENT
73.996	0.4564	41.5151	4301	61.998	1.2361	43.5478	5801
75.993	5.7320	41.4434	4302	63.999	7.4169	43.4718	5802
77.991	10.8956	41.3856	4303	66.000	13.4638	43.4033	5803
79.991	15.9845	41.3278	4304	68.001	19.4467	43.3348	5804
81.992	21.0491	41.2691	4305				
				59.999	0.5011	43.8506	4701
73.996	0.4815	41.5190	10601	61.998	6.8268	43.7690	4702
75.993	5.7438	41.4471	10602	63.999	13.0040	43.6990	4703
77.991	10.8992	41.3889	10603	66.000	19.1342	43.6290	4704
79.991	15.9884	41.3308	10604				
81.992	21.0641	41.2719	10605	59.999	0.5139	43.8515	11601
				61.998	6.8361	43.7701	11602
73.996	0.5595	41.5209	10401	63.999	13.0332	43.7001	11603
75.993	5.8320	41.4497	10402	66.000	19.1699	43.6300	11604
77.991	10.9885	41.3914	10403				
79.991	16.0681	41.3333	10404	58.002	1.5812	44.2038	4801
81.992	21.1332	41.2744	10405	59.999	8.0600	44.1262	4802
				61.998	14.4207	44.0544	4803
71.998	0.8278	41.8772	6101	63.999	20.7475	43.9822	4804
73.996	6.2426	41.8068	6102				
75.993	11.5453	41.7467	6103	56.005	1.4389	44.5234	6201
77.991	16.7781	41.6867	6104	58.002	8.0623	44.4435	6202
79.991	21.9801	41.6259	6105	59.999	14.5535	44.3698	6203
				61.998	21.0109	44.2954	6204
70.000	0.4372	42.1990	10701				
71.998	6.0066	42.1238	10702	56.005	4.8597	44.6360	12201
73.996	11.4350	42.0624	10703	58.002	11.4722	44.5618	12202
75.993	16.8041	42.0012	10704	59.999	18.0087	44.4877	12203
77.991	22.1545	41.9388	10705	61.998	24.5494	44.4105	12204
70.000	0.4585	42.1996	4101	53.610	1.3141		12001
71.998	6.0266	42.1252	4102	53.709	2.2350		12002
73.996	11.4696	42.0640	4103	53.809	3.3653		12003
75.993	16.8566	42.0028	4104	53.909	4.3943		12004
77.991	22.2168	41.9406	4105	54.009	5.0652	44.9752	12005
				54.109	5.4099	44.9721	12006
68.001	0.8566	42.5531	6001	54.209	5.7571	44.9682	12007
70.000	6.5671	42.4798	6002	55.007	8.5036	44.9373	12008
71.998	12.1452	42.4164	6003	56.005	11.8881	44.8989	12009
73.996	17.6633	42.3532	6004	58.002	18.5846	44.8229	12010
75.993	23.1733	42.2884	6005				
				53.909	4.3901		12101
66.000	0.8508	42.8768	5901	54.009	5.4182		12102
68.001	6.7269	42.8021	5902	54.109	6.4487		12103
70.000	12.4546	42.7372	5903	54.209	7.4857		12104
71.998	18.1377	42.6724	5904	54.309	8.5322		12105
73.996	23.7925	42.6058	5905	54.408	9.5650		12106
				54.508	10.6060		12107
63.999	0.9133	43.2041	5401	54.608	11.6500		12108
66.000	6.9342	43.1279	5402	54.708	12.6979		12109
68.001	12.8113	43.0610	5403	54.808	13.7392		12110
70.000	18.6316	42.9941	5404	54.907	14.7088	45.1393	12111
				55.007	15.0681	45.1353	12112
63.999	1.2662	43.2189	4901	55.107	15.4160	45.1315	12113
66.000	7.2989	43.1446	4902	56.005	18.4968	45.0966	12114
68.001	13.1924	43.0780	4903				
70.000	19.0462	43.0114	4904				

LE 15. THERMODYNAMIC PROPERTIES OF FLUORINE

ON THE SATURATION BOUNDARY

number of significant figures given in this table is not
fied on the basis of the uncertainty of the data, but is
ented to maintain internal consistency.

TABLE 15. THERMODYNAMIC PROPERTIES OF FLUORINE ON THE SATURATION BOUNDARY*

TEMPERATURE (IPTS 1968) K	PRESSURE MN/m²	DENSITY MOL/L	ISOTHERM DERIVATIVE J/MOL	ISOCHORE DERIVATIVE MN/m²-K	INTERNAL ENERGY J/MOL	ENTHALPY J/MOL	ENTROPY J/MOL-K	C_v J / MOL - K	C_p	VELOCITY OF SOUND M/S
53.4811	0.00025	44.8623	26740	4.308	-6025.1	-6025.1	60.86	36.51	54.96	1029
53.4811	0.00025	0.00057	444		1108.0	1552.5	201.90	20.80	29.12	128
54	0.00030	44.7808	26701	4.284	-5996.8	-5996.8	61.39	36.42	54.94	1029
54	0.00030	0.00066	449		1118.7	1567.6	200.84	20.80	29.12	129
56	0.00054	44.4648	26280	4.189	-5887.4	-5887.3	63.38	36.07	54.99	1027
56	0.00054	0.00115	465		1160.1	1625.5	196.95	20.80	29.13	131
58	0.00093	44.1463	25976	4.090	-5777.3	-5777.3	65.31	35.70	54.86	1025
58	0.00093	0.00193	481		1201.4	1683.2	193.39	20.81	29.15	133
60	0.00155	43.8251	25738	3.988	-5666.7	-5666.7	67.18	35.31	54.62	1024
60	0.00155	0.00311	498		1242.5	1740.7	190.13	20.81	29.17	135
62	0.00249	43.5012	25561	3.884	-5555.7	-5555.6	69.00	34.92	54.25	1022
62	0.00249	0.00484	514		1283.5	1798.0	187.14	20.82	29.20	138
64	0.00387	43.1744	24827	3.778	-5444.3	-5444.2	70.77	34.51	54.25	1013
64	0.00387	0.00729	529		1324.1	1854.8	184.38	20.84	29.24	140
66	0.00584	42.8446	23657	3.672	-5332.5	-5332.4	72.49	34.10	54.60	998
66	0.00584	0.01068	545	0.0001	1364.5	1911.2	181.84	20.85	29.29	142
68	0.00858	42.5117	22808	3.566	-5220.4	-5220.2	74.17	33.69	54.67	987
68	0.00858	0.01525	560	0.0001	1404.6	1967.1	179.49	20.88	29.36	144
70	0.01230	42.1756	21592	3.460	-5108.0	-5107.7	75.80	33.28	55.10	970
70	0.01230	0.02128	574	0.0002	1444.2	2022.3	177.31	20.90	29.44	146
72	0.01725	41.8360	20430	3.354	-4995.3	-4994.9	77.38	32.87	55.52	953
72	0.01725	0.02907	588	0.0002	1483.3	2076.8	175.28	20.94	29.54	148
74	0.02372	41.4929	20177	3.249	-4882.3	-4881.8	78.93	32.46	54.95	948
74	0.02372	0.03897	602	0.0003	1521.8	2130.3	173.40	20.98	29.66	150
76	0.03201	41.1460	19216	3.145	-4769.1	-4768.3	80.44	32.07	55.17	933
76	0.03201	0.05135	615	0.0004	1559.6	2182.9	171.64	21.03	29.80	151
78	0.04246	40.7952	17992	3.041	-4655.6	-4654.5	81.91	31.68	55.78	913
78	0.04246	0.06660	627	0.0006	1596.7	2234.3	169.99	21.08	29.97	153
80	0.05547	40.4401	17116	2.939	-4541.8	-4540.4	83.36	31.31	56.00	898
80	0.05547	0.08515	638	0.0007	1633.0	2284.5	168.44	21.15	30.17	155
82	0.07144	40.0807	16310	2.837	-4427.6	-4425.8	84.77	30.96	56.15	882
82	0.07144	0.10745	648	0.0009	1668.3	2333.2	166.99	21.22	30.40	156
84	0.09081	39.7165	15452	2.737	-4313.2	-4310.9	86.14	30.62	56.43	866
84	0.09081	0.13397	657	0.0011	1702.7	2380.5	165.62	21.31	30.67	158
86	0.11403	39.3472	14495	2.637	-4198.3	-4195.4	87.50	30.30	56.96	847
86	0.11403	0.16523	665	0.0014	1735.9	2426.1	164.32	21.41	30.97	159
88	0.14161	38.9727	13638	2.539	-4083.0	-4079.4	88.82	30.01	57.39	829
88	0.14161	0.20174	672	0.0017	1768.0	2469.9	163.09	21.51	31.32	160
90	0.17403	38.5924	12602	2.442	-3967.3	-3962.8	90.12	29.73	58.33	807
90	0.17403	0.24407	677	0.0021	1798.8	2511.8	161.93	21.64	31.71	162
92	0.21184	38.2060	12013	2.347	-3851.1	-3845.5	91.40	29.48	58.37	791
92	0.21184	0.29280	682	0.0026	1828.2	2551.7	160.81	21.77	32.15	163
94	0.25558	37.8131	11449	2.253	-3734.3	-3727.5	92.66	29.24	58.39	776
94	0.25558	0.34857	684	0.0031	1856.2	2589.4	159.75	21.92	32.65	164
96	0.30581	37.4131	10621	2.161	-3616.8	-3608.6	93.89	29.03	59.18	755
96	0.30581	0.41203	686	0.0037	1882.6	2624.8	158.73	22.08	33.20	165
98	0.36309	37.0055	10013	2.070	-3498.6	-3488.8	95.11	28.85	59.46	737
98	0.36309	0.48389	685	0.0044	1907.3	2657.7	157.74	22.26	33.82	166
100	0.42802	36.5897	9296	1.981	-3379.7	-3368.0	96.32	28.68	60.21	717
100	0.42802	0.56491	684	0.0051	1930.3	2688.0	156.80	22.46	34.52	166

DYNAMIC PROPERTIES OF FLUORINE ON THE SATURATION BOUNDARY — CONTINUED

ISOTHERM DERIVATIVE J/MOL	ISOCHORE DERIVATIVE MN/m²-K	INTERNAL ENERGY J/MOL	ENTHALPY J/MOL	ENTROPY J/MOL-K	C_V J / MOL - K	C_p J / MOL - K	VELOCITY OF SOUND M/S
8751	-1.894	-3259.9	-3246.0	97.50	28.52	60.49	699
680	0.0060	1951.4	2715.5	155.88	22.66	35.29	167
8206	1.809	-3139.1	-3122.7	98.68	28.39	60.86	680
674	0.0070	1970.5	2740.2	154.99	22.89	36.16	167
7486	1.725	-3017.2	-2998.1	99.84	28.27	62.12	658
667	0.0082	1987.5	2761.7	154.13	23.13	37.12	168
6862	1.643	-2894.1	-2871.8	100.99	28.17	63.21	637
658	0.0095	2002.2	2779.9	153.28	23.39	38.20	168
6291	1.564	-2769.7	-2743.9	102.14	28.08	64.29	616
647	0.0110	2014.6	2794.6	152.45	23.66	39.42	168
5886	1.485	-2643.9	-2614.0	103.28	28.00	64.58	598
633	0.0126	2024.3	2805.6	151.63	23.95	40.79	168
5247	1.409	-2516.4	-2482.0	104.41	27.93	66.64	574
618	0.0145	2031.2	2812.6	150.82	24.26	42.34	168
4784	1.334	-2387.1	-2347.7	105.54	27.87	67.83	553
600	0.0167	2035.0	2815.3	150.02	24.59	44.11	168
4252	1.260	-2255.8	-2210.7	106.67	27.83	70.02	531
580	0.0191	2035.5	2813.2	149.22	24.94	46.15	168
3817	1.188	-2122.1	-2070.7	107.80	27.80	71.75	509
557	0.0219	2032.4	2806.0	148.42	25.32	48.52	168
3429	1.117	-1985.9	-1927.4	108.93	27.79	73.39	488
532	0.0251	2025.1	2793.1	147.61	25.71	51.30	167
2982	1.046	-1846.7	-1780.3	110.07	27.81	76.49	465
504	0.0287	2013.2	2773.8	146.79	26.14	54.62	167
2608	0.977	-1704.0	-1628.8	111.23	27.84	79.28	442
473	0.0329	1996.0	2747.3	145.95	26.59	58.67	166
2245	0.908	-1557.1	-1472.0	112.40	27.91	82.81	419
439	0.0377	1972.5	2712.4	145.08	27.08	63.71	165
1862	0.840	-1405.4	-1309.1	113.59	28.01	88.49	393
402	0.0434	1941.7	2667.6	144.18	27.62	70.19	164
1526	0.772	-1247.6	-1138.6	114.82	28.14	95.10	368
360	0.0501	1901.8	2610.8	143.22	28.22	78.87	163
1211	0.705	-1081.9	-958.5	116.09	28.26	104.31	343
314	0.0582	1850.3	2538.7	142.19	28.91	91.15	161
933	0.634	-906.6	-766.5	117.42	28.44	115.43	316
263	0.0683	1783.0	2446.2	141.04	29.71	110.06	160
635	0.561	-718.0	-558.0	118.85	29.90	140.66	280
210	0.0795	1692.6	2324.4	139.73	30.96	136.11	156
384	0.489	-504.5	-320.1	120.45	31.07	188.62	248
152	0.0957	1569.0	2161.3	138.17	33.10	190.40	152
186	0.402	-246.6	-29.5	122.39	32.15	293.49	211
85	0.121	1385.7	1923.7	136.14	35.45	350.23	149
7	0.282	182.9	463.7	125.69	42.95	5191.48	145
12	0.173	973.7	1406.6	132.24	41.44	2537.44	140
0	0.227	562.3	907.6	128.74		∞	
0	0.227	562.3	907.6	128.74		∞	

ERATURE REFERS TO THE LIQUID PHASE.

TABLE 16. THERMODYNAMIC PROPERTIES OF FLUORINE

The number of significant figures given in this table is not justified on the basis of the uncertainty of the data, but is presented to maintain internal consistency.

TABLE 16. THERMODYNAMIC PROPERTIES OF FLUORINE

0.01 MN/m² ISOBAR

TEMPERATURE (IPTS 1968) K	DENSITY MOL/L	ISOTHERM DERIVATIVE J/MOL	ISOCHORE DERIVATIVE MN/m²-K	INTERNAL ENERGY J/MOL	ENTHALPY J/MOL	ENTROPY J/MOL-K	C_v J / MOL - K	C_p J / MOL - K	VELOCITY OF SOUND M/S
* 53.482	44.8775	26743	4.3103	-6027.0	-6026.8	60.83	36.51	54.95	1029
54	44.7927	26705	4.3099	-5998.4	-5998.2	61.36	36.42	55.14	1032
56	44.4653	26284	4.2967	-5887.6	-5887.4	63.38	36.07	55.98	1036
58	44.1409	25951	4.2042	-5776.8	-5776.6	65.32	35.70	55.98	1035
60	43.8254	25739	4.1275	-5666.9	-5666.7	67.18	35.31	55.99	1036
62	43.5015	25561	4.0613	-5555.9	-5555.7	69.00	34.92	56.06	1039
64	43.1678	24864	3.9690	-5443.5	-5443.3	70.79	34.50	56.26	1033
66	42.8327	23699	3.8247	-5331.0	-5330.7	72.52	34.08	56.29	1015
68	42.5119	22808	3.6371	-5220.4	-5220.2	74.17	33.69	55.51	995
* 68.835	42.3718	22346	3.5216	-5173.5	-5173.3	74.85	33.52	54.80	981
* 68.835	0.0175724	566	0.0001	1421.2	1990.2	178.56	20.89	29.39	145
70	0.0172751	576	0.0001	1445.6	2024.5	179.05	20.88	29.37	146
72	0.0167879	593	0.0001	1487.5	2083.2	179.88	20.87	29.35	148
74	0.0163278	610	0.0001	1529.4	2141.9	180.68	20.87	29.33	150
76	0.0158926	627	0.0001	1571.3	2200.5	181.46	20.86	29.31	152
78	0.0154802	643	0.0001	1613.2	2259.1	182.22	20.86	29.30	154
80	0.0150890	660	0.0001	1655.0	2317.7	182.97	20.85	29.28	156
82	0.0147172	677	0.0001	1696.8	2376.3	183.69	20.85	29.27	158
84	0.0143635	694	0.0001	1738.6	2434.8	184.39	20.84	29.26	160
86	0.0140265	711	0.0001	1780.4	2493.3	185.08	20.84	29.25	162
88	0.0137051	728	0.0001	1822.2	2551.8	185.76	20.84	29.24	164
90	0.0133982	744	0.0001	1863.9	2610.3	186.41	20.84	29.23	166
92	0.0131048	761	0.0001	1905.7	2668.7	187.05	20.83	29.22	168
94	0.0128241	778	0.0001	1947.4	2727.2	187.68	20.83	29.22	169
96	0.0125552	795	0.0001	1989.1	2785.6	188.30	20.83	29.21	171
98	0.0122975	812	0.0001	2030.9	2844.0	188.90	20.83	29.21	173
100	0.0120501	828	0.0001	2072.6	2902.4	189.49	20.83	29.20	-175
102	0.0118126	845	0.0001	2114.3	2960.8	190.07	20.83	29.20	177
104	0.0115843	862	0.0001	2156.0	3019.2	190.64	20.83	29.19	178
106	0.0113647	879	0.0001	2197.7	3077.6	191.19	20.83	29.19	180
108	0.0111533	895	0.0001	2239.4	3136.0	191.74	20.83	29.19	182
110	0.0109497	912	0.0001	2281.1	3194.4	192.27	20.83	29.19	183
112	0.0107533	929	0.0001	2322.8	3252.7	192.80	20.83	29.18	185
114	0.0105640	945	0.0001	2364.5	3311.1	193.32	20.83	29.18	187
116	0.0103812	962	0.0001	2406.2	3369.5	193.82	20.83	29.18	188
118	0.0102046	979	0.0001	2447.9	3427.8	194.32	20.83	29.18	190
120	0.0100340	996	0.0001	2489.6	3486.2	194.81	20.83	29.18	192
122	0.0098690	1012	0.0001	2531.3	3544.6	195.29	20.84	29.18	193
124	0.0097093	1029	0.0001	2573.0	3603.0	195.77	20.84	29.19	195
126	0.0095548	1046	0.0001	2614.7	3661.3	196.24	20.84	29.19	196
128	0.0094051	1062	0.0001	2656.4	3719.7	196.70	20.85	29.19	198
130	0.0092600	1079	0.0001	2698.2	3778.1	197.15	20.85	29.19	199
132	0.0091193	1096	0.0001	2739.9	3836.5	197.59	20.86	29.20	201
134	0.0089829	1112	0.0001	2781.7	3894.9	198.03	20.86	29.20	202
136	0.0088505	1129	0.0001	2823.4	3953.3	198.47	20.87	29.21	204
138	0.0087219	1146	0.0001	2865.2	4011.7	198.89	20.88	29.21	205
140	0.0085971	1162	0.0001	2907.0	4070.2	199.31	20.88	29.22	207
142	0.0084757	1179	0.0001	2948.8	4128.6	199.73	20.89	29.23	208
144	0.0083578	1196	0.0001	2990.6	4187.1	200.14	20.90	29.23	210
146	0.0082431	1212	0.0001	3032.4	4245.5	200.54	20.91	29.24	211
148	0.0081315	1229	0.0001	3074.2	4304.0	200.94	20.92	29.25	213
150	0.0080229	1246	0.0001	3116.1	4362.5	201.33	20.93	29.26	214
152	0.0079171	1262	0.0001	3158.0	4421.1	201.72	20.94	29.27	216
154	0.0078141	1279	0.0001	3199.9	4479.6	202.10	20.95	29.28	217
156	0.0077138	1296	0.0001	3241.8	4538.2	202.48	20.96	29.29	218
158	0.0076160	1312	0.0001	3283.8	4596.8	202.85	20.97	29.30	220
160	0.0075207	1329	0.0001	3325.7	4655.4	203.22	20.99	29.32	221
165	0.0072925	1371	0.0001	3430.4	4802.1	204.12	21.02	29.35	224
170	0.0070777	1412	0.0001	3536.0	4948.9	205.00	21.06	29.39	228
175	0.0068753	1454	0.0001	3641.5	5096.0	205.85	21.11	29.43	231
180	0.0066841	1496	0.0001	3747.2	5243.3	206.68	21.15	29.48	234
185	0.0065032	1537	0.0001	3853.1	5390.8	207.49	21.20	29.53	237
190	0.0063320	1579	0.0001	3959.3	5538.6	208.28	21.26	29.58	240
195	0.0061695	1620	0.0001	4065.7	5686.6	209.05	21.32	29.64	244
200	0.0060151	1662	0.0001	4172.5	5835.0	209.80	21.38	29.71	247
205	0.0058683	1704		4279.6	5983.7	210.53	21.45	29.77	249
210	0.0057285	1745		4387.0	6132.7	211.25	21.52	29.84	252
215	0.0055952	1787		4494.8	6282.1	211.95	21.59	29.91	255
220	0.0054679	1829		4603.0	6431.8	212.64	21.66	29.98	258
225	0.0053463	1870		4711.5	6581.9	213.32	21.74	30.06	261
230	0.0052300	1912		4820.4	6732.4	213.98	21.82	30.14	264
235	0.0051187	1953		4929.7	6883.3	214.63	21.90	30.22	266
240	0.0050120	1995		5039.4	7034.6	215.27	21.98	30.30	269
245	0.0049097	2037		5149.5	7186.3	215.89	22.07	30.39	272
250	0.0048115	2078		5260.1	7338.5	216.51	22.15	30.47	274
255	0.0047171	2120		5371.1	7491.1	217.11	22.24	30.56	277
260	0.0046263	2161		5482.5	7644.1	217.70	22.33	30.65	279
265	0.0045390	2203		5594.4	7797.5	218.29	22.42	30.74	282
270	0.0044549	2245		5706.7	7951.4	218.86	22.51	30.83	284
275	0.0043739	2286		5819.5	8105.8	219.43	22.60	30.92	287
280	0.0042958	2328		5932.7	8260.6	219.99	22.69	31.01	289
285	0.0042204	2369		6046.4	8415.8	220.54	22.78	31.10	292
290	0.0041476	2411		6160.5	8571.6	221.08	22.87	31.19	294
295	0.0040773	2453		6275.1	8727.7	221.61	22.96	31.28	297
300	0.0040093	2494		6390.1	8884.3	222.14	23.05	31.37	299

* TWO-PHASE BOUNDARY

TABLE 16. THERMODYNAMIC PROPERTIES OF FLUORINE

0.02 MN/m² ISOBAR

TEMPERATURE (IPTS 1968) K	DENSITY MOL/L	ISOTHERM DERIVATIVE J/MOL	ISOCHORE DERIVATIVE MN/m²-K	INTERNAL ENERGY J/MOL	ENTHALPY J/MOL	ENTROPY J/MOL-K	C_v J / MOL - K	C_p J / MOL - K	VELOCITY OF SOUND M/S
* 53.483	44.8777	26748	4.3103	-6027.3	-6026.8	60.83	36.51	54.95	1029
54	44.7930	26709	4.3100	-5998.7	-5998.2	61.36	36.42	55.14	1032
56	44.4657	26286	4.2989	-5887.9	-5887.5	63.38	36.07	55.98	1036
58	44.1413	25953	4.2044	-5777.1	-5776.6	65.32	35.70	55.98	1035
60	43.8257	25740	4.1275	-5667.2	-5666.8	67.18	35.31	55.99	1036
62	43.5019	25560	4.0613	-5556.2	-5555.7	69.00	34.92	56.06	1039
64	43.1682	24862	3.9960	-5443.8	-5443.3	70.79	34.50	56.26	1033
66	42.8331	23697	3.8248	-5331.3	-5330.8	72.52	34.08	56.29	1015
68	42.5124	22808	3.6373	-5220.7	-5220.3	74.17	33.69	55.52	995
70	42.1819	21605	3.5182	-5109.0	-5108.6	75.78	33.28	55.82	977
72	41.8322	20432	3.4154	-4994.8	-4994.3	77.39	32.86	56.35	960
* 72.913	41.6798	19912	3.3063	-4943.7	-4943.3	78.09	32.68	55.72	945
* 72.913	0.0333111	595	0.0003	1500.9	2101.3	174.41	20.95	29.59	149
74	0.0328081	604	0.0003	1523.9	2133.5	174.84	20.95	29.57	150
76	0.0319220	621	0.0003	1566.1	2192.6	175.63	20.93	29.53	152
78	0.0310836	638	0.0003	1608.2	2251.6	176.40	20.92	29.50	154
80	0.0302891	655	0.0003	1650.3	2310.6	177.14	20.91	29.47	156
82	0.0295351	673	0.0002	1692.3	2369.5	177.87	20.90	29.44	158
84	0.0288183	690	0.0002	1734.3	2428.3	178.58	20.90	29.42	160
86	0.0281361	707	0.0002	1776.3	2487.1	179.27	20.89	29.39	162
88	0.0274860	724	0.0002	1818.3	2545.9	179.95	20.88	29.38	164
90	0.0268657	741	0.0002	1860.2	2604.6	180.61	20.88	29.36	166
92	0.0262732	758	0.0002	1902.1	2663.3	181.25	20.87	29.34	167
94	0.0257066	774	0.0002	1944.0	2722.0	181.88	20.87	29.33	169
96	0.0251642	791	0.0002	1985.9	2780.7	182.50	20.86	29.31	171
98	0.0246444	808	0.0002	2027.7	2839.3	183.11	20.86	29.30	173
100	0.0241460	825	0.0002	2069.6	2897.9	183.70	20.86	29.29	175
102	0.0236675	842	0.0002	2111.4	2956.4	184.28	20.85	29.28	176
104	0.0232077	859	0.0002	2153.2	3015.0	184.85	20.85	29.27	178
106	0.0227657	876	0.0002	2195.0	3073.5	185.40	20.85	29.27	180
108	0.0223403	893	0.0002	2236.8	3132.1	185.95	20.85	29.26	182
110	0.0219307	909	0.0002	2278.6	3190.6	186.49	20.85	29.25	183
112	0.0215359	926	0.0002	2320.4	3249.1	187.01	20.85	29.25	185
114	0.0211552	943	0.0002	2362.2	3307.6	187.53	20.85	29.24	187
116	0.0207878	960	0.0002	2403.9	3366.0	188.04	20.85	29.24	188
118	0.0204330	977	0.0002	2445.7	3424.5	188.54	20.85	29.24	190
120	0.0200902	993	0.0002	2487.5	3483.0	189.03	20.85	29.23	191
122	0.0197588	1010	0.0002	2529.2	3541.4	189.51	20.85	29.23	193
124	0.0194382	1027	0.0002	2571.0	3599.9	189.99	20.85	29.23	195
126	0.0191278	1044	0.0002	2612.8	3658.4	190.46	20.86	29.23	196
128	0.0188273	1060	0.0002	2654.6	3716.8	190.92	20.86	29.23	198
130	0.0185362	1077	0.0002	2696.3	3775.3	191.37	20.86	29.23	199
132	0.0182539	1094	0.0002	2738.1	3833.8	191.82	20.87	29.24	201
134	0.0179801	1111	0.0001	2779.9	3892.3	192.26	20.87	29.24	202
136	0.0177145	1127	0.0001	2821.7	3950.7	192.69	20.88	29.24	204
138	0.0174566	1144	0.0001	2863.5	4009.2	193.12	20.88	29.25	205
140	0.0172062	1161	0.0001	2905.3	4067.7	193.54	20.89	29.25	207
142	0.0169629	1177	0.0001	2947.2	4126.2	193.95	20.90	29.26	208
144	0.0167264	1194	0.0001	2989.0	4184.7	194.36	20.91	29.26	210
146	0.0164964	1211	0.0001	3030.9	4243.3	194.77	20.91	29.27	211
148	0.0162726	1228	0.0001	3072.8	4301.8	195.16	20.92	29.28	213
150	0.0160549	1244	0.0001	3114.7	4360.4	195.56	20.93	29.29	214
152	0.0158429	1261	0.0001	3156.6	4419.0	195.95	20.94	29.29	215
154	0.0156365	1278	0.0001	3198.5	4477.6	196.33	20.95	29.30	217
156	0.0154354	1294	0.0001	3240.5	4536.2	196.71	20.97	29.31	218
158	0.0152394	1311	0.0001	3282.4	4594.8	197.08	20.98	29.33	220
160	0.0150484	1328	0.0001	3324.4	4653.5	197.45	20.99	29.34	221
165	0.0145911	1370	0.0001	3429.6	4800.3	198.35	21.03	29.37	224
170	0.0141609	1411	0.0001	3534.9	4947.2	199.23	21.07	29.41	228
175	0.0137554	1453	0.0001	3640.4	5094.3	200.08	21.11	29.45	231
180	0.0133725	1495	0.0001	3746.3	5241.7	200.91	21.15	29.49	235
185	0.0130103	1536	0.0001	3852.0	5389.3	201.72	21.21	29.54	237
190	0.0126673	1578	0.0001	3958.3	5537.1	202.51	21.26	29.60	240
195	0.0123420	1620	0.0001	4064.8	5685.3	203.28	21.32	29.66	243
200	0.0120329	1661	0.0001	4171.6	5833.7	204.03	21.38	29.72	246
205	0.0117390	1703	0.0001	4278.7	5982.4	204.77	21.45	29.78	249
210	0.0114591	1745	0.0001	4386.2	6131.5	205.48	21.52	29.85	252
215	0.0111923	1786	0.0001	4494.0	6280.9	206.19	21.59	29.92	255
220	0.0109376	1828	0.0001	4602.2	6430.7	206.88	21.66	29.99	258
225	0.0106942	1870	0.0001	4710.7	6580.9	207.55	21.74	30.07	261
230	0.0104615	1911	0.0001	4819.6	6731.4	208.21	21.82	30.15	264
235	0.0102387	1953	0.0001	4929.0	6882.3	208.86	21.90	30.23	266
240	0.0100252	1995	0.0001	5038.7	7033.7	209.50	21.98	30.31	269
245	0.0098204	2036	0.0001	5148.9	7185.4	210.13	22.07	30.39	272
250	0.0096238	2078	0.0001	5259.4	7337.6	210.74	22.15	30.48	274
255	0.0094350	2119	0.0001	5370.5	7490.2	211.34	22.24	30.57	277
260	0.0092534	2161	0.0001	5481.9	7643.3	211.94	22.33	30.65	279
265	0.0090787	2203	0.0001	5593.8	7796.8	212.52	22.42	30.74	282
270	0.0089104	2244	0.0001	5706.1	7950.7	213.10	22.51	30.83	284
275	0.0087483	2286	0.0001	5818.9	8105.1	213.67	22.60	30.92	287
280	0.0085920	2327	0.0001	5932.2	8259.9	214.22	22.69	31.01	289
285	0.0084412	2369	0.0001	6045.9	8415.2	214.77	22.78	31.10	292
290	0.0082956	2411	0.0001	6160.0	8570.9	215.31	22.87	31.19	294
295	0.0081549	2452	0.0001	6274.6	8727.1	215.85	22.96	31.28	297
300	0.0080189	2494	0.0001	6389.7	8883.8	216.38	23.05	31.37	299

* TWO-PHASE BOUNDARY

TABLE 16. THERMODYNAMIC PROPERTIES OF FLUORINE

0.04 MN/m² ISOBAR

TEMPERATURE (IPTS 1968) K	DENSITY MOL/L	ISOTHERM DERIVATIVE J/MOL	ISOCHORE DERIVATIVE MN/m²-K	INTERNAL ENERGY J/MOL	ENTHALPY J/MOL	ENTROPY J/MOL-K	C_v J / MOL - K	C_p J / MOL - K	VELOCITY OF SOUND M/S
* 53.485	44.8781	26756	4.3104	-6027.7	-6026.8	60.83	36.50	54.95	1030
54	44.7938	26717	4.3101	-5999.2	-5998.3	61.36	36.42	55.13	1032
56	44.4665	26292	4.2992	-5888.4	-5887.5	63.38	36.07	55.98	1036
58	44.1421	25956	4.2047	-5777.6	-5776.7	65.32	35.70	55.98	1035
60	43.8265	25744	4.1276	-5667.8	-5666.9	57.18	35.31	55.99	1036
62	43.5027	25559	4.0611	-5558.8	-5555.8	69.00	34.92	56.06	1039
64	43.1690	24857	3.9689	-5444.4	-5443.4	70.79	34.50	56.27	1033
66	42.8340	23694	3.8251	-5331.8	-5330.9	72.52	34.09	56.30	1015
68	42.5133	22809	3.6377	-5221.3	-5220.4	74.16	33.69	55.52	995
70	42.1828	21607	3.5184	-5109.6	-5108.7	75.78	33.28	55.82	977
72	41.8331	20431	3.4154	-4995.4	-4994.5	77.39	32.86	56.35	960
74	41.4937	20174	3.3355	-4882.8	-4881.9	78.93	32.46	56.17	958
76	41.1465	19215	3.1955	-4769.3	-4768.4	80.44	32.07	55.92	939
* 77.568	40.8714	18078	3.0638	-4680.1	-4679.2	81.60	31.76	55.88	915
* 77.568	0.0630412	624	0.0005	1588.8	2223.3	170.34	21.07	29.93	153
78	0.0626746	628	0.0005	1598.0	2236.2	170.50	21.06	29.92	153
80	0.0610351	646	0.0005	1640.6	2296.0	171.26	21.04	29.85	155
82	0.0594828	663	0.0005	1683.2	2355.6	172.00	21.02	29.80	157
84	0.0580105	681	0.0005	1725.6	2415.2	172.71	21.00	29.74	159
86	0.0566119	698	0.0005	1768.1	2474.6	173.41	20.99	29.70	161
88	0.0552814	715	0.0005	1810.4	2534.0	174.09	20.97	29.65	163
90	0.0540139	733	0.0005	1852.7	2593.2	174.76	20.96	29.62	165
92	0.0528048	750	0.0004	1894.9	2652.4	175.41	20.95	29.58	167
94	0.0516501	767	0.0004	1937.1	2711.6	176.05	20.94	29.55	169
96	0.0505461	785	0.0004	1979.3	2770.6	176.67	20.93	29.52	171
98	0.0494893	802	0.0004	2021.4	2829.7	177.28	20.92	29.50	-172
100	0.0484769	819	0.0004	2063.5	2888.6	177.87	20.91	29.47	174
102	0.0475058	836	0.0004	2105.6	2947.6	178.46	20.91	29.45	176
104	0.0465736	853	0.0004	2147.6	3006.4	179.03	20.90	29.43	178
106	0.0456780	870	0.0004	2189.6	3065.3	179.59	20.90	29.42	180
108	0.0448167	887	0.0004	2231.6	3124.1	180.14	20.89	29.40	181
110	0.0439878	904	0.0004	2273.6	3182.9	180.68	20.89	29.39	183
112	0.0431895	921	0.0004	2315.5	3241.7	181.21	20.88	29.37	185
114	0.0424201	938	0.0004	2357.5	3300.4	181.73	20.88	29.36	186
116	0.0416780	955	0.0003	2399.4	3359.1	182.24	20.88	29.35	188
118	0.0409617	972	0.0003	2441.3	3417.8	182.74	20.88	29.34	190
120	0.0402699	989	0.0003	2483.2	3476.5	183.23	20.88	29.34	191
122	0.0396014	1006	0.0003	2525.1	3535.2	183.72	20.88	29.33	193
124	0.0389549	1023	0.0003	2567.0	3593.8	184.19	20.88	29.32	194
126	0.0383295	1040	0.0003	2608.9	3652.5	184.66	20.88	29.32	196
128	0.0377240	1056	0.0003	2650.8	3711.1	185.13	20.88	29.32	198
130	0.0371375	1073	0.0003	2692.6	3769.7	185.58	20.88	29.31	199
132	0.0365691	1090	0.0003	2734.5	3828.3	186.03	20.89	29.31	201
134	0.0360180	1107	0.0003	2776.4	3887.0	186.47	20.89	29.31	202
136	0.0354834	1124	0.0003	2818.3	3945.6	186.90	20.90	29.31	204
138	0.0349646	1141	0.0003	2860.2	4004.2	187.33	20.90	29.31	205
140	0.0344609	1157	0.0003	2902.1	4062.8	187.75	20.91	29.31	207
142	0.0339715	1174	0.0003	2944.0	4121.5	188.17	20.91	29.32	208
144	0.0334960	1191	0.0003	2985.9	4180.1	188.58	20.92	29.32	210
146	0.0330336	1208	0.0003	3027.9	4238.8	188.98	20.93	29.33	211
148	0.0325840	1225	0.0003	3069.8	4297.4	189.38	20.94	29.33	212
150	0.0321465	1241	0.0003	3111.8	4356.1	189.78	20.95	29.34	214
152	0.0317206	1258	0.0003	3153.7	4414.8	190.16	20.95	29.34	215
154	0.0313060	1275	0.0003	3195.7	4473.4	190.55	20.97	29.35	217
156	0.0309021	1292	0.0003	3237.8	4532.2	190.93	20.98	29.36	218
158	0.0305086	1309	0.0003	3279.8	4590.9	191.30	20.99	29.37	220
160	0.0301250	1325	0.0003	3321.8	4649.6	191.67	21.00	29.38	221
165	0.0292071	1367	0.0002	3427.1	4796.6	192.57	21.03	29.41	224
170	0.0283438	1409	0.0002	3532.5	4943.7	193.45	21.07	29.44	228
175	0.0275302	1451	0.0002	3638.1	5091.0	194.31	21.12	29.48	231
180	0.0267622	1493	0.0002	3743.9	5238.6	195.14	21.16	29.52	234
185	0.0260360	1535	0.0002	3850.0	5386.3	195.95	21.21	29.57	237
190	0.0253483	1576	0.0002	3956.3	5534.3	196.74	21.27	29.62	240
195	0.0246961	1618	0.0002	4062.9	5682.6	197.51	21.33	29.68	243
200	0.0240767	1660	0.0002	4169.7	5831.1	198.26	21.39	29.74	246
205	0.0234877	1702	0.0002	4276.9	5980.0	198.99	21.45	29.80	249
210	0.0229269	1743	0.0002	4384.5	6129.1	199.71	21.52	29.87	252
215	0.0223923	1785	0.0002	4492.3	6278.6	200.42	21.59	29.94	255
220	0.0218822	1827	0.0002	4600.5	6428.5	201.11	21.67	30.01	258
225	0.0213948	1869	0.0002	4709.1	6578.8	201.78	21.74	30.09	261
230	0.0209287	1910	0.0002	4818.1	6729.4	202.44	21.82	30.16	264
235	0.0204825	1952	0.0002	4927.5	6880.4	203.09	21.90	30.24	266
240	0.0200550	1994	0.0002	5037.3	7031.8	203.73	21.99	30.32	269
245	0.0196449	2035	0.0002	5147.5	7183.6	204.36	22.07	30.41	272
250	0.0192514	2077	0.0002	5258.1	7335.9	204.97	22.16	30.49	274
255	0.0188733	2119	0.0002	5369.2	7488.6	205.58	22.24	30.58	277
260	0.0185098	2160	0.0002	5480.7	7641.7	206.17	22.33	30.67	279
265	0.0181600	2202	0.0002	5592.6	7795.2	206.76	22.42	30.75	282
270	0.0178233	2244	0.0001	5705.0	7949.2	207.33	22.51	30.84	284
275	0.0174988	2285	0.0001	5817.8	8103.6	207.90	22.60	30.93	287
280	0.0171860	2327	0.0001	5931.1	8258.5	208.46	22.69	31.02	289
285	0.0168841	2369	0.0001	6044.8	8413.9	209.01	22.78	31.11	292
290	0.0165927	2410	0.0001	6159.0	8569.7	209.55	22.87	31.20	294
295	0.0163112	2452	0.0001	6273.6	8725.9	210.08	22.96	31.29	297
300	0.0160391	2494	0.0001	6388.7	8882.6	210.61	23.06	31.38	299

* TWO-PHASE BOUNDARY

TABLE 16. THERMODYNAMIC PROPERTIES OF FLUORINE

0.06 MN/m² ISOBAR

TEMPERATURE (IPTS 1968) K	DENSITY MOL/L	ISOTHERM DERIVATIVE J/MOL	ISOCHORE DERIVATIVE MN/m²-K	INTERNAL ENERGY J/MOL	ENTHALPY J/MOL	ENTROPY J/MOL-K	C_v J / MOL - K	C_p J / MOL - K	VELOCITY OF SOUND M/S
* 53.487	44.8786	26764	4.3105	-6028.1	-6026.8	60.83	36.50	54.94	1030
54	44.7945	26724	4.3102	-5999.8	-5998.4	61.36	36.42	55.13	1032
56	44.4672	26297	4.2995	-5889.0	-5887.6	63.37	36.07	55.98	1036
58	44.1428	25960	4.2050	-5778.2	-5776.8	65.32	35.70	55.97	1035
60	43.8273	25747	4.1278	-5668.3	-5667.0	67.18	35.32	55.99	1036
62	43.5034	25558	4.0610	-5557.3	-5555.9	69.00	34.92	56.06	1039
64	43.1698	24853	3.9688	-5444.9	-5443.5	70.78	34.50	56.27	1033
66	42.8348	23691	3.8253	-5332.4	-5331.0	72.51	34.09	56.30	1015
68	42.5141	22810	3.6381	-5221.9	-5220.5	74.16	33.69	55.52	995
70	42.1838	21608	3.5187	-5110.2	-5108.8	75.78	33.29	55.82	977
72	41.8341	20431	3.4155	-4996.0	-4994.6	77.39	32.87	56.36	960
74	41.4947	20171	3.3356	-4883.5	-4882.0	78.93	32.47	56.17	958
76	41.1475	19213	3.1957	-4770.0	-4768.5	80.44	32.07	55.93	939
78	40.7962	17992	3.0796	-4656.1	-4654.7	81.91	31.68	56.39	918
80	40.4404	17116	2.9789	-4541.9	-4540.4	83.36	31.31	56.67	903
* 80.609	40.3312	15961	2.9079	-4507.1	-4505.6	83.79	31.20	57.46	879
* 80.609	0.0915190	641	0.0008	1643.9	2299.5	167.99	21.17	30.24	155
82	0.0898642	653	0.0008	1673.8	2341.5	168.51	21.15	30.17	157
84	0.0875939	671	0.0007	1716.8	2401.7	169.24	21.12	30.09	159
86	0.0854418	689	0.0007	1759.6	2461.9	159.94	21.09	30.02	161
88	0.0833981	707	0.0007	1802.4	2521.8	170.63	21.07	29.95	163
90	0.0814545	725	0.0007	1845.0	2581.7	171.30	21.05	29.89	165
92	0.0796033	743	0.0007	1887.6	2641.4	171.96	21.03	29.83	166
94	0.0778378	760	0.0007	1930.2	2701.0	172.60	21.01	29.78	168
96	0.0761519	778	0.0006	1972.6	2760.5	173.23	21.00	29.74	170
98	0.0745400	795	0.0006	2015.0	2820.0	173.84	20.99	29.70	172
100	0.0729971	812	0.0006	2057.4	2879.3	174.44	20.97	29.66	174
102	0.0715188	830	0.0006	2099.7	2938.6	175.03	20.96	29.63	176
104	0.0701009	847	0.0006	2141.9	2997.8	175.60	20.95	29.60	177
106	0.0687397	864	0.0006	2184.1	3057.0	176.17	20.94	29.57	179
108	0.0674316	882	0.0006	2226.3	3116.1	176.72	20.94	29.55	181
110	0.0661737	899	0.0006	2268.5	3175.2	177.26	20.93	29.52	183
112	0.0649628	916	0.0005	2310.6	3234.2	177.79	20.92	29.50	184
114	0.0637964	933	0.0005	2352.7	3293.2	178.31	20.92	29.48	186
116	0.0626720	950	0.0005	2394.8	3352.2	178.83	20.91	29.47	188
118	0.0615874	967	0.0005	2436.9	3411.1	179.33	20.91	29.45	189
120	0.0605403	984	0.0005	2478.9	3470.0	179.83	20.91	29.44	191
122	0.0595288	1001	0.0005	2520.9	3528.8	180.31	20.91	29.43	193
124	0.0585512	1019	0.0005	2562.9	3587.7	180.79	20.91	29.42	194
126	0.0576056	1036	0.0005	2604.9	3646.5	181.26	20.91	29.41	196
128	0.0566905	1053	0.0005	2646.9	3705.3	181.72	20.91	29.40	197
130	0.0558045	1070	0.0005	2688.9	3764.1	182.18	20.91	29.39	199
132	0.0549461	1086	0.0005	2730.9	3822.9	182.63	20.91	29.39	200
134	0.0541141	1103	0.0005	2772.9	3881.7	183.07	20.91	29.38	202
136	0.0533072	1120	0.0004	2814.9	3940.4	183.51	20.91	29.38	204
138	0.0525243	1137	0.0004	2856.9	3999.2	183.93	20.92	29.38	205
140	0.0517643	1154	0.0004	2898.8	4057.9	184.36	20.92	29.38	207
142	0.0510262	1171	0.0004	2940.8	4116.7	184.77	20.93	29.38	208
144	0.0503091	1188	0.0004	2982.8	4175.5	185.19	20.93	29.38	209
146	0.0496121	1205	0.0004	3024.8	4234.2	185.59	20.94	29.38	211
148	0.0489343	1222	0.0004	3066.8	4293.0	185.99	20.95	29.38	212
150	0.0482750	1239	0.0004	3108.9	4351.8	186.38	20.96	29.39	214
152	0.0476333	1255	0.0004	3150.9	4410.5	186.77	20.97	29.39	215
154	0.0470086	1272	0.0004	3193.0	4469.3	187.16	20.98	29.40	217
156	0.0464002	1289	0.0004	3235.0	4528.1	187.54	20.99	29.41	218
158	0.0458075	1306	0.0004	3277.1	4587.0	187.91	21.00	29.41	219
160	0.0452299	1323	0.0004	3319.2	4645.8	188.28	21.01	29.42	221
165	0.0438480	1365	0.0004	3424.6	4793.0	189.19	21.04	29.45	224
170	0.0425486	1407	0.0004	3530.1	4940.3	190.07	21.08	29.48	228
175	0.0413245	1449	0.0003	3635.8	5087.8	190.92	21.12	29.51	231
180	0.0401691	1491	0.0003	3741.7	5235.4	191.75	21.17	29.56	234
185	0.0390770	1533	0.0003	3847.9	5383.3	192.57	21.22	29.60	237
190	0.0380430	1575	0.0003	3954.3	5531.4	193.36	21.27	29.65	240
195	0.0370625	1616	0.0003	4060.9	5679.8	194.13	21.33	29.70	243
200	0.0361314	1658	0.0003	4167.9	5828.5	194.88	21.39	29.76	246
205	0.0352462	1700	0.0003	4275.2	5977.5	195.61	21.46	29.82	249
210	0.0344035	1742	0.0003	4382.7	6126.7	196.33	21.53	29.89	252
215	0.0336003	1784	0.0003	4490.7	6276.4	197.04	21.60	29.96	255
220	0.0328338	1826	0.0003	4598.9	6426.3	197.73	21.67	30.03	258
225	0.0321017	1867	0.0003	4707.6	6576.7	198.40	21.75	30.10	261
230	0.0314015	1909	0.0003	4816.6	6727.4	199.07	21.83	30.18	264
235	0.0307314	1951	0.0003	4926.0	6878.4	199.72	21.91	30.26	266
240	0.0300893	1993	0.0003	5035.9	7029.9	200.35	21.99	30.34	269
245	0.0294736	2034	0.0002	5146.1	7181.8	200.98	22.07	30.42	272
250	0.0288826	2076	0.0002	5256.8	7334.2	201.60	22.16	30.51	274
255	0.0283149	2118	0.0002	5367.9	7486.9	202.20	22.25	30.59	277
260	0.0277692	2160	0.0002	5479.4	7640.1	202.80	22.33	30.68	279
265	0.0272441	2201	0.0002	5591.4	7793.7	203.38	22.42	30.77	282
270	0.0267386	2243	0.0002	5703.8	7947.7	203.96	22.51	30.85	284
275	0.0262515	2285	0.0002	5816.6	8102.2	204.52	22.60	30.94	287
280	0.0257819	2326	0.0002	5929.4	8257.2	205.08	22.69	31.03	289
285	0.0253288	2368	0.0002	6043.7	8412.6	205.63	22.79	31.12	292
290	0.0248914	2410	0.0002	6157.9	8568.4	206.17	22.88	31.21	294
295	0.0244688	2451	0.0002	6272.6	8724.7	206.71	22.97	31.30	297
300	0.0240604	2493	0.0002	6387.7	8881.4	207.23	23.06	31.39	299

* TWO-PHASE BOUNDARY

TABLE 16. THERMODYNAMIC PROPERTIES OF FLUORINE

0.08 MN/m² ISOBAR

TEMPERATURE (IPTS 1968) K	DENSITY MOL/L	ISOTHERM DERIVATIVE J/MOL	ISOCHORE DERIVATIVE MN/m²-K	INTERNAL ENERGY J/MOL	ENTHALPY J/MOL	ENTROPY J/MOL-K	C_v J / MOL - K	C_p	VELOCITY OF SOUND M/S
* 53.489	44.8790	26773	4.3106	-6028.5	-6026.7	60.83	36.50	54.93	1030
54	44.7953	26732	4.3103	-6000.3	-5998.5	61.36	36.42	55.12	1032
56	44.4680	26302	4.2998	-5889.5	-5887.7	63.37	36.07	55.98	1036
58	44.1436	25963	4.2053	-5778.7	-5776.9	65.32	35.70	55.97	1035
60	43.8281	25750	4.1279	-5668.9	-5667.1	67.18	35.32	55.98	1036
62	43.5042	25558	4.0609	-5557.9	-5556.0	69.00	34.92	56.06	1039
64	43.1706	24848	3.9687	-5445.5	-5443.7	70.78	34.51	56.27	1033
66	42.8357	23688	3.8255	-5333.0	-5331.1	72.51	34.09	56.31	1015
68	42.5150	22811	3.6385	-5222.5	-5220.6	74.16	33.69	55.53	995
70	42.1847	21610	3.5189	-5110.8	-5108.9	75.78	33.29	55.82	977
72	41.8351	20430	3.4155	-4996.7	-4994.7	77.38	32.87	56.36	960
74	41.4957	20168	3.3357	-4884.1	-4882.2	78.93	32.47	56.18	958
76	41.1486	19212	3.1958	-4770.6	-4768.7	80.44	32.07	55.93	939
78	40.7973	17992	3.0797	-4656.8	-4654.8	81.91	31.69	56.39	918
80	40.4416	17116	2.9790	-4542.6	-4540.6	83.35	31.31	56.68	903
82	40.0715	16323	2.8636	-4426.5	-4424.5	84.78	30.94	56.60	886
82.930	39.9119	15700	2.7904	-4374.4	-4372.4	85.41	30.80	56.62	872
* 82.930	0.119231	652	0.0010	1684.4	2355.4	166.34	21.26	30.52	157
84	0.117587	662	0.0010	1707.7	2388.0	166.73	21.24	30.46	158
86	0.114641	681	0.0010	1751.0	2448.8	167.45	21.20	30.35	160
88	0.111849	699	0.0010	1794.2	2509.5	168.15	21.17	30.26	162
90	0.109198	717	0.0009	1837.3	2569.9	168.82	21.14	30.17	164
92	0.106678	735	0.0009	1880.2	2630.1	169.49	21.12	30.10	166
94	0.104277	753	0.0009	1923.1	2690.3	170.13	21.09	30.03	168
96	0.101988	771	0.0009	1965.9	2750.3	170.77	21.07	29.96	170
98	0.0998017	788	0.0008	2008.5	2810.1	171.38	21.05	29.91	172
100	0.0977114	806	0.0008	2051.2	2869.9	171.99	21.03	29.86	174
102	0.0957105	824	0.0008	2093.7	2929.6	172.58	21.02	29.81	175
104	0.0937931	841	0.0008	2136.2	2989.1	173.16	21.01	29.77	177
106	0.0919538	859	0.0008	2178.6	3048.6	173.72	20.99	29.73	179
108	0.0901877	876	0.0008	2221.0	3108.1	174.28	20.98	29.70	181
110	0.0884903	894	0.0007	2263.4	3167.4	174.82	20.97	29.66	182
112	0.0868577	911	0.0007	2305.7	3226.7	175.36	20.96	29.64	184
114	0.0852859	928	0.0007	2348.0	3286.0	175.88	20.96	29.61	186
116	0.0837715	946	0.0007	2390.2	3345.2	176.40	20.95	29.59	187
118	0.0823113	963	0.0007	2432.4	3404.3	176.90	20.94	29.56	189
120	0.0809025	980	0.0007	2474.6	3463.4	177.40	20.94	29.54	191
122	0.0795421	997	0.0007	2516.7	3522.5	177.89	20.93	29.53	192
124	0.0782278	1014	0.0007	2558.9	3581.5	178.37	20.93	29.51	194
126	0.0769571	1031	0.0006	2601.0	3640.5	178.84	20.93	29.50	196
128	0.0757278	1049	0.0006	2643.1	3699.5	179.30	20.93	29.49	197
130	0.0745379	1066	0.0006	2685.2	3758.5	179.76	20.93	29.47	199
132	0.0733855	1083	0.0006	2727.3	3817.4	180.21	20.93	29.47	200
134	0.0722689	1100	0.0006	2769.4	3876.3	180.65	20.93	29.46	202
136	0.0711862	1117	0.0006	2811.4	3935.3	181.09	20.93	29.45	203
138	0.0701360	1134	0.0006	2853.5	3994.2	181.52	20.94	29.45	205
140	0.0691169	1151	0.0006	2895.6	4053.0	181.94	20.94	29.44	206
142	0.0681273	1168	0.0006	2937.6	4111.9	182.36	20.94	29.44	208
144	0.0671661	1185	0.0006	2979.7	4170.8	182.77	20.95	29.44	209
146	0.0662320	1202	0.0006	3021.8	4229.7	183.18	20.96	29.44	211
148	0.0653238	1219	0.0005	3063.9	4288.5	183.58	20.96	29.44	212
150	0.0644405	1236	0.0005	3106.0	4347.4	183.97	20.97	29.44	214
152	0.0635811	1253	0.0005	3148.1	4406.3	184.36	20.98	29.44	215
154	0.0627445	1270	0.0005	3190.2	4465.2	184.75	20.99	29.45	217
156	0.0619299	1287	0.0005	3232.3	4524.1	185.13	21.00	29.45	218
158	0.0611364	1303	0.0005	3274.5	4583.0	185.50	21.01	29.46	219
160	0.0603632	1320	0.0005	3316.6	4641.9	185.87	21.02	29.47	221
165	0.0585139	1363	0.0005	3422.1	4789.3	186.78	21.05	29.49	224
170	0.0567756	1405	0.0005	3527.8	4936.8	187.66	21.09	29.51	227
175	0.0551383	1447	0.0005	3633.6	5084.5	188.52	21.13	29.59	231
180	0.0535935	1489	0.0004	3739.6	5232.3	189.35	21.17	29.59	234
185	0.0521334	1531	0.0004	3845.8	5380.3	190.16	21.22	29.63	237
190	0.0507513	1573	0.0004	3952.3	5528.6	190.95	21.28	29.68	240
195	0.0494410	1615	0.0004	4059.0	5677.1	191.72	21.34	29.73	243
200	0.0481971	1657	0.0004	4166.0	5825.9	192.48	21.40	29.79	246
205	0.0470145	1699	0.0004	4273.4	5975.0	193.21	21.46	29.85	249
210	0.0458888	1741	0.0004	4381.0	6124.4	193.93	21.53	29.91	252
215	0.0448161	1783	0.0004	4489.0	6274.1	194.64	21.60	29.98	255
220	0.0437925	1824	0.0004	4597.3	6424.1	195.33	21.67	30.05	258
225	0.0428148	1866	0.0004	4706.0	6574.5	196.00	21.75	30.12	261
230	0.0418801	1908	0.0003	4815.1	6725.3	196.67	21.83	30.20	264
235	0.0409854	1950	0.0003	4924.6	6876.5	197.32	21.91	30.27	266
240	0.0401282	1992	0.0003	5034.5	7028.1	197.96	21.99	30.35	269
245	0.0393063	2034	0.0003	5144.7	7180.0	198.58	22.08	30.44	272
250	0.0385175	2075	0.0003	5255.5	7332.4	199.20	22.16	30.52	274
255	0.0377599	2117	0.0003	5366.6	7485.2	199.80	22.25	30.60	277
260	0.0370315	2159	0.0003	5478.2	7638.5	200.40	22.34	30.69	279
265	0.0363308	2201	0.0003	5590.2	7792.1	200.98	22.43	30.78	282
270	0.0356562	2242	0.0003	5702.6	7946.3	201.56	22.52	30.87	284
275	0.0350063	2284	0.0003	5815.5	8100.8	202.13	22.61	30.95	287
280	0.0343797	2326	0.0003	5928.8	8255.8	202.69	22.70	31.04	289
285	0.0337751	2368	0.0003	6042.6	8411.2	203.24	22.79	31.13	292
290	0.0331916	2409	0.0003	6156.9	8567.1	203.78	22.88	31.22	294
295	0.0326279	2451	0.0003	6271.6	8723.5	204.31	22.97	31.31	297
300	0.0320830	2493	0.0003	6386.7	8880.2	204.84	23.06	31.40	299

* TWO-PHASE BOUNDARY

TABLE 16. THERMODYNAMIC PROPERTIES OF FLUORINE

0.101325 MN/m² (1 ATM) ISOBAR

TEMPERATURE (IPTS 1968) K	DENSITY MOL/L	ISOTHERM DERIVATIVE J/MOL	ISOCHORE DERIVATIVE MN/m²-K	INTERNAL ENERGY J/MOL	ENTHALPY J/MOL	ENTROPY J/MOL-K	C_v J / MOL - K	C_p J / MOL - K	VELOCITY OF SOUND M/S
* 53.491	44.8794	26782	4.3107	-6029.0	-6026.7	60.83	36.50	54.93	1030
54	44.7961	26741	4.3104	-6000.9	-5998.6	61.36	36.42	55.11	1032
56	44.4688	26308	4.3001	-5890.1	-5887.8	63.37	36.07	55.97	1037
58	44.1444	25967	4.2056	-5779.3	-5777.0	65.31	35.70	55.97	1035
60	43.8289	25754	4.1280	-5669.5	-5667.2	67.18	35.32	55.98	1037
62	43.5051	25557	4.0608	-5558.5	-5556.2	69.00	34.93	56.06	1039
64	43.1714	24843	3.9686	-5446.1	-5443.8	70.78	34.51	56.28	1033
66	42.8366	23685	3.8258	-5333.6	-5331.3	72.51	34.09	56.32	1015
68	42.5159	22812	3.6389	-5223.1	-5220.8	74.16	33.69	55.53	995
70	42.1857	21612	3.5191	-5111.5	-5109.1	75.78	33.29	55.82	977
72	41.8361	20430	3.4155	-4997.3	-4994.9	77.38	32.87	56.36	960
74	41.4968	20165	3.3358	-4884.7	-4882.3	78.92	32.47	56.19	958
76	41.1497	19210	3.1960	-4771.3	-4768.8	80.43	32.07	55.54	939
78	40.7985	17992	3.0798	-4657.5	-4655.0	81.91	31.69	56.39	918
80	40.4428	17116	2.9971	-4543.3	-4540.8	83.35	31.32	56.68	903
82	40.0728	16321	2.8638	-4427.2	-4424.7	84.78	30.94	56.60	886
84	39.7172	15452	2.7488	-4313.5	-4311.0	86.14	30.62	56.66	867
* 84.950	39.5418	14790	2.6894	-4258.7	-4256.1	86.79	30.47	57.04	854
* 84.950	0.148191	661	0.0013	1718.6	2402.4	165.00	21.35	30.81	158
86	0.146200	671	0.0013	1741.6	2434.7	165.37	21.33	30.73	160
88	0.142568	690	0.0012	1785.3	2496.0	166.08	21.28	30.61	162
90	0.139127	708	0.0012	1828.8	2557.1	166.76	21.25	30.49	164
92	0.135860	727	0.0012	1872.2	2618.0	167.43	21.21	30.39	166
94	0.132755	745	0.0011	1915.4	2678.7	168.09	21.18	30.30	167
96	0.129797	763	0.0011	1958.5	2739.2	168.72	21.15	30.21	169
98	0.126976	781	0.0011	2001.6	2799.5	169.35	21.13	30.14	171
100	0.124282	799	0.0011	2044.5	2859.8	169.95	21.10	30.07	173
102	0.121706	817	0.0010	2087.3	2919.8	170.55	21.08	30.01	175
104	0.119240	835	0.0010	2130.0	2979.8	171.13	21.06	29.96	177
106	0.116877	853	0.0010	2172.7	3039.7	171.70	21.05	29.90	179
108	0.114610	870	0.0010	2215.3	3099.4	172.26	21.03	29.86	180
110	0.112432	888	0.0009	2257.9	3159.1	172.81	21.02	29.82	182
112	0.110339	905	0.0009	2300.4	3218.7	173.34	21.01	29.78	184
114	0.108326	923	0.0009	2342.8	3278.2	173.87	21.00	29.74	186
116	0.106387	940	0.0009	2385.3	3337.7	174.39	20.99	29.71	187
118	0.104518	958	0.0009	2427.6	3397.1	174.90	20.98	29.68	189
120	0.102716	975	0.0009	2470.0	3456.4	175.39	20.97	29.66	191
122	0.100977	993	0.0008	2512.3	3515.7	175.88	20.97	29.63	192
124	0.0992975	1010	0.0008	2554.5	3575.0	176.37	20.96	29.61	194
126	0.0976745	1027	0.0008	2596.8	3634.2	176.84	20.96	29.59	195
128	0.0961049	1044	0.0008	2639.0	3693.3	177.31	20.95	29.58	197
130	0.0945863	1062	0.0008	2681.2	3752.5	177.76	20.95	29.56	199
132	0.0931159	1079	0.0008	2723.4	3811.6	178.22	20.95	29.55	200
134	0.0916916	1096	0.0008	2765.6	3870.7	178.66	20.95	29.54	202
136	0.0903111	1113	0.0008	2807.8	3929.7	179.10	20.95	29.53	203
138	0.0889724	1130	0.0007	2849.9	3988.8	179.53	20.95	29.52	205
140	0.0876736	1147	0.0007	2892.1	4047.8	179.95	20.96	29.51	206
142	0.0864128	1165	0.0007	2934.2	4106.8	180.37	20.96	29.51	208
144	0.0851885	1182	0.0007	2976.4	4165.8	180.78	20.97	29.50	209
146	0.0839989	1199	0.0007	3018.6	4224.8	181.19	20.97	29.50	211
148	0.0828426	1216	0.0007	3060.7	4283.8	181.59	20.98	29.50	212
150	0.0817182	1233	0.0007	3102.9	4342.8	181.99	20.98	29.50	214
152	0.0806243	1250	0.0007	3145.0	4401.8	182.38	20.99	29.50	215
154	0.0795598	1267	0.0007	3187.2	4460.8	182.76	21.00	29.50	216
156	0.0785234	1284	0.0007	3229.4	4519.8	183.14	21.01	29.50	218
158	0.0775141	1301	0.0006	3271.6	4578.8	183.52	21.02	29.51	219
160	0.0765307	1318	0.0006	3313.8	4637.8	183.89	21.03	29.51	221
165	0.0741792	1360	0.0006	3419.5	4785.4	184.80	21.06	29.53	224
170	0.0719695	1402	0.0006	3525.2	4933.1	185.68	21.10	29.55	227
175	0.0698889	1445	0.0006	3631.1	5081.0	186.54	21.14	29.58	231
180	0.0679263	1487	0.0006	3737.3	5229.0	187.37	21.18	29.62	234
185	0.0660719	1529	0.0006	3843.6	5377.1	188.19	21.23	29.66	237
190	0.0643168	1571	0.0005	3950.1	5525.6	188.98	21.28	29.70	240
195	0.0626532	1613	0.0005	4057.0	5674.2	189.75	21.34	29.76	243
200	0.0610741	1655	0.0005	4164.1	5823.1	190.50	21.40	29.81	246
205	0.0595732	1697	0.0005	4271.5	5972.3	191.24	21.47	29.87	249
210	0.0581447	1739	0.0005	4379.2	6121.8	191.96	21.53	29.93	252
215	0.0567835	1781	0.0005	4487.2	6271.6	192.67	21.60	30.00	255
220	0.0554849	1823	0.0005	4595.6	6421.8	193.36	21.68	30.07	258
225	0.0542447	1865	0.0005	4704.4	6572.3	194.03	21.75	30.14	261
230	0.0530590	1907	0.0004	4813.5	6723.2	194.70	21.83	30.21	264
235	0.0519243	1949	0.0004	4923.0	6874.4	195.35	21.91	30.29	266
240	0.0508373	1991	0.0004	5033.0	7026.1	195.98	22.00	30.37	269
245	0.0497951	2033	0.0004	5143.3	7178.1	196.61	22.08	30.45	272
250	0.0487949	2075	0.0004	5254.0	7330.6	197.23	22.17	30.53	274
255	0.0478342	2116	0.0004	5365.2	7483.5	197.83	22.25	30.62	277
260	0.0469108	2158	0.0004	5476.8	7636.8	198.43	22.34	30.70	279
265	0.0460225	2200	0.0004	5588.9	7790.5	199.01	22.43	30.79	282
270	0.0451673	2242	0.0004	5701.4	7944.7	199.59	22.52	30.88	284
275	0.0443435	2284	0.0004	5814.3	8099.3	200.16	22.61	30.97	287
280	0.0435492	2325	0.0004	5927.7	8254.3	200.72	22.70	31.05	289
285	0.0427830	2367	0.0004	6041.5	8409.8	201.27	22.79	31.14	292
290	0.0420434	2409	0.0004	6155.8	8565.8	201.81	22.88	31.23	294
295	0.0413289	2451	0.0003	6270.5	8722.2	202.34	22.97	31.32	297
300	0.0406384	2492	0.0003	6385.7	8879.0	202.87	23.06	31.41	299

* TWO-PHASE BOUNDARY

TABLE 16. THERMODYNAMIC PROPERTIES OF FLUORINE

15 MN/m² ISOBAR

JRE ;8)	DENSITY MOL/L	ISOTHERM DERIVATIVE J/MOL	ISOCHORE DERIVATIVE MN/m²-K	INTERNAL ENERGY J/MOL	ENTHALPY J/MOL	ENTROPY J/MOL-K	c_v J / MOL - K	c_p	VELOCITY OF SOUND M/S
›96	44.8805	26802	4.3109	-6030.0	-6026.7	60.83	36.50	54.92	1030
	44.7979	26760	4.3106	-6002.2	-5998.8	61.35	36.42	55.10	1032
	44.4707	26321	4.3009	-5891.4	-5888.1	63.37	36.07	55.97	1037
	44.1463	25976	4.2063	-5780.7	-5777.3	65.31	35.70	55.97	1035
	43.8308	25762	4.1283	-5670.8	-5667.4	67.17	35.32	55.98	1037
	43.5070	25554	4.0605	-5559.9	-5556.4	68.99	34.93	56.06	1039
	43.1734	24833	3.9683	-5447.5	-5444.0	70.78	34.51	56.29	1032
	42.8386	23678	3.8263	-5335.0	-5331.5	72.51	34.09	56.33	1015
	42.5181	22813	3.6399	-5224.6	-5221.0	74.15	33.70	55.54	995
	42.1879	21617	3.5196	-5112.9	-5109.4	75.77	33.29	55.83	977
	41.8385	20428	3.4157	-4998.8	-4995.2	77.38	32.87	56.36	960
	41.4992	20157	3.3361	-4886.3	-4882.7	78.92	32.48	56.20	958
	41.1522	19206	3.1965	-4772.8	-4769.2	80.43	32.08	55.95	939
	40.8012	17992	3.0800	-4659.1	-4655.4	81.90	31.69	56.40	918
	40.4457	17117	2.9794	-4544.9	-4541.2	83.35	31.32	56.68	903
	40.0757	16317	2.8641	-4428.9	-4425.1	84.77	30.95	56.62	886
	39.7203	15455	2.7493	-4315.2	-4311.4	86.14	30.63	56.66	868
	39.3497	14498	2.6287	-4199.6	-4195.8	87.49	30.30	56.78	845
	38.9733	13639	2.5104	-4083.4	-4079.5	88.82	30.01	56.77	824
;49	38.8689	13183	2.5124	-4051.3	-4047.5	89.18	29.93	57.99	820
;49	0.212750	674	0.0018	1776.6	2481.6	162.77	21.55	31.42	161
	0.208860	688	0.0018	1808.9	2527.1	163.28	21.50	31.28	162
	0.203753	707	0.0018	1853.3	2589.5	163.96	21.44	31.11	164
	0.198917	726	0.0017	1897.5	2651.6	164.63	21.39	30.96	166
	0.194327	746	0.0017	1941.5	2713.4	165.28	21.34	30.82	168
	0.189965	764	0.0016	1985.3	2774.9	165.92	21.30	30.70	170
	0.185811	783	0.0016	2028.9	2836.2	166.53	21.27	30.59	172
	0.181850	802	0.0016	2072.4	2897.2	167.14	21.23	30.49	174
	0.178066	820	0.0015	2115.8	2958.1	167.73	21.20	30.40	176
	0.174448	838	0.0015	2159.0	3018.9	168.31	21.17	30.32	178
	0.170984	857	0.0015	2202.1	3079.4	168.87	21.15	30.25	180
	0.167663	875	0.0014	2245.2	3139.8	169.43	21.13	30.18	181
	0.164476	893	0.0014	2288.2	3200.1	169.97	21.11	30.12	183
	0.161415	911	0.0014	2331.0	3260.3	170.50	21.09	30.06	185
	0.158472	929	0.0013	2373.9	3320.4	171.03	21.07	30.01	187
	0.155639	946	0.0013	2416.6	3380.4	171.54	21.06	29.97	188
	0.152910	964	0.0013	2459.3	3440.3	172.04	21.05	29.92	190
	0.150280	982	0.0013	2501.9	3500.1	172.54	21.04	29.89	192
	0.147742	1000	0.0012	2544.5	3559.8	173.02	21.03	29.85	193
	0.145292	1017	0.0012	2587.1	3619.5	173.50	21.02	29.82	195
	0.142925	1035	0.0012	2629.6	3679.1	173.97	21.01	29.79	196
	0.140637	1052	0.0012	2672.1	3738.6	174.43	21.01	29.76	198
	0.138423	1070	0.0012	2714.5	3798.2	174.89	21.00	29.74	200
	0.136280	1087	0.0011	2756.9	3857.6	175.33	21.00	29.72	201
	0.134205	1105	0.0011	2799.3	3917.0	175.77	21.00	29.70	203
	0.132193	1122	0.0011	2841.7	3976.4	176.21	21.00	29.68	204
	0.130243	1139	0.0011	2884.1	4035.8	176.63	21.00	29.67	206
	0.128351	1157	0.0011	2926.4	4095.1	177.05	21.00	29.66	207
	0.126515	1174	0.0011	2968.8	4154.4	177.47	21.00	29.65	209
	0.124732	1191	0.0010	3011.1	4213.7	177.88	21.00	29.64	210
	0.122999	1209	0.0010	3053.4	4273.0	178.28	21.01	29.63	212
	0.121315	1226	0.0010	3095.8	4332.2	178.68	21.01	29.62	213
	0.119678	1243	0.0010	3138.1	4391.5	179.07	21.02	29.62	215
	0.118085	1260	0.0010	3180.4	4450.7	179.46	21.03	29.62	216
	0.116535	1277	0.0010	3222.8	4509.9	179.84	21.04	29.62	218
	0.115026	1294	0.0010	3265.1	4569.2	180.22	21.04	29.62	219
	0.113556	1312	0.0010	3307.5	4628.4	180.59	21.05	29.62	220
	0.110043	1354	0.0009	3413.4	4776.5	181.50	21.08	29.63	224
	0.106745	1397	0.0009	3519.4	4924.7	182.39	21.12	29.64	227
	0.103641	1440	0.0009	3625.6	5072.9	183.25	21.15	29.66	231
	0.100716	1482	0.0008	3732.0	5221.3	184.08	21.20	29.69	234
	0.0979527	1525	0.0008	3838.5	5369.9	184.90	21.24	29.73	237
	0.0953391	1567	0.0008	3945.5	5518.6	185.69	21.30	29.77	240
	0.0928627	1609	0.0008	4052.3	5667.6	186.46	21.35	29.82	243
	0.0905131	1652	0.0008	4159.5	5816.8	187.22	21.41	29.87	246
	0.0882806	1694	0.0007	4267.1	5966.2	187.96	21.48	29.92	249
	0.0861565	1736	0.0007	4375.0	6116.0	188.68	21.54	29.98	252
	0.0841331	1778	0.0007	4483.2	6266.0	189.38	21.61	30.04	255
	0.0822034	1820	0.0007	4591.7	6416.4	190.08	21.69	30.11	258
	0.0803608	1862	0.0007	4700.6	6567.2	190.75	21.76	30.18	261
	0.0785996	1905	0.0007	4809.8	6718.2	191.42	21.84	30.25	263
	0.0769145	1947	0.0006	4919.5	6869.7	192.07	21.92	30.33	266
	0.0753006	1989	0.0006	5029.5	7021.5	192.71	22.00	30.41	269
	0.0737534	2031	0.0006	5139.9	7173.7	193.34	22.09	30.49	272
	0.0722690	2073	0.0006	5250.8	7326.4	193.95	22.17	30.57	274
	0.0708434	2115	0.0006	5362.1	7479.4	194.56	22.26	30.65	277
	0.0694733	2156	0.0006	5473.8	7632.9	195.15	22.35	30.73	279
	0.0681555	2198	0.0006	5585.9	7786.8	195.74	22.44	30.82	282
	0.0668870	2240	0.0006	5698.5	7941.1	196.32	22.53	30.91	284
	0.0656650	2282	0.0005	5811.5	8095.8	196.89	22.62	30.99	287
	0.0644872	2324	0.0005	5925.0	8251.0	197.45	22.71	31.08	289
	0.0633510	2366	0.0005	6038.9	8406.6	198.00	22.80	31.17	292
	0.0622544	2408	0.0005	6153.2	8562.7	198.54	22.89	31.26	294
	0.0611952	2450	0.0005	6268.0	8719.2	199.07	22.98	31.35	297
	0.0601716	2491	0.0005	6383.3	8876.2	199.60	23.07	31.44	299

-PHASE BOUNDARY

TABLE 16. THERMODYNAMIC PROPERTIES OF FLUORINE

0.2 MN/m² ISOBAR

TEMPERATURE (IPTS 1968) K	DENSITY MOL/L	ISOTHERM DERIVATIVE J/MOL	ISOCHORE DERIVATIVE MN/m²-K	INTERNAL ENERGY J/MOL	ENTHALPY J/MOL	ENTROPY J/MOL-K	C_v J / MOL - K	C_p	VELOCITY OF SOUND M/S
* 53.501	44.8815	26823	4.3112	-6031.1	-6026.6	60.83	36.50	54.90	1030
54	44.7998	26780	4.3108	-6003.5	-5999.0	61.35	36.41	55.09	1033
56	44.4726	26334	4.3016	-5892.8	-5888.3	63.36	36.07	55.96	1037
58	44.1482	25985	4.2071	-5782.0	-5777.5	65.31	35.70	55.97	1035
60	43.8327	25771	4.1286	-5672.2	-5667.7	67.17	35.32	55.97	1037
62	43.5089	25552	4.0602	-5561.3	-5556.7	68.99	34.93	56.06	1039
64	43.1754	24822	3.9681	-5449.0	-5444.3	70.77	34.52	56.30	1032
66	42.8407	23670	3.8269	-5336.5	-5331.8	72.50	34.10	56.35	1015
68	42.5203	22815	3.6409	-5226.1	-5221.3	74.15	33.70	55.55	995
70	42.1902	21622	3.5201	-5114.5	-5109.7	75.77	33.29	55.83	977
72	41.8410	20427	3.4158	-5000.4	-4995.6	77.37	32.87	56.37	960
74	41.5017	20149	3.3361	-4887.8	-4883.0	78.91	32.48	56.21	958
76	41.1548	19202	3.1970	-4774.4	-4769.6	80.42	32.08	55.97	939
78	40.8040	17992	3.0802	-4660.7	-4655.8	81.90	31.70	56.40	918
80	40.4486	17117	2.9797	-4546.6	-4541.6	83.34	31.33	56.69	903
82	40.0788	16313	2.8645	-4430.6	-4425.6	84.77	30.95	56.63	886
84	39.7236	15459	2.7499	-4316.9	-4311.9	86.13	30.63	56.67	868
86	39.3532	14503	2.6294	-4201.4	-4196.3	87.49	30.31	56.78	846
88	38.9770	13664	2.5107	-4085.2	-4080.0	88.81	30.01	56.77	824
90	38.5945	12608	2.4430	-3968.3	-3963.1	90.12	29.73	58.34	807
* 91.405	38.3217	12196	2.3751	-3885.7	-3880.5	91.02	29.55	58.34	796
* 91.405	0.277595	681	0.0024	1819.6	2540.1	161.14	21.73	32.01	162
92	0.275490	687	0.0024	1833.1	2559.1	161.35	21.71	31.94	163
94	0.268677	707	0.0023	1878.4	2622.7	162.03	21.63	31.71	165
96	0.262240	727	0.0023	1923.3	2686.0	162.70	21.56	31.51	167
98	0.256144	747	0.0022	1968.0	2748.8	163.34	21.50	31.33	169
100	0.250358	766	0.0022	2012.4	2811.3	163.97	21.45	31.17	171
102	0.244856	786	0.0021	2056.7	2873.5	164.59	21.40	31.02	173
104	0.239616	805	0.0021	2100.7	2935.4	165.19	21.35	30.89	175
106	0.234617	824	0.0020	2144.6	2997.0	165.78	21.31	30.77	177
108	0.229841	842	0.0020	2188.3	3058.5	166.35	21.28	30.67	179
110	0.225272	861	0.0019	2231.9	3119.7	166.91	21.25	30.57	181
112	0.220895	880	0.0019	2275.4	3180.8	167.46	21.22	30.49	182
114	0.216698	898	0.0018	2318.7	3241.7	168.00	21.19	30.41	184
116	0.212668	916	0.0018	2362.0	3302.4	168.53	21.17	30.33	186
118	0.208796	935	0.0018	2405.1	3363.0	169.05	21.15	30.27	188
120	0.205071	953	0.0017	2448.2	3423.5	169.56	21.13	30.21	189
122	0.201484	971	0.0017	2491.2	3483.8	170.06	21.11	30.15	191
124	0.198027	989	0.0017	2534.1	3544.1	170.55	21.10	30.10	193
126	0.194694	1007	0.0016	2577.0	3604.3	171.03	21.09	30.06	194
128	0.191476	1025	0.0016	2619.8	3664.3	171.50	21.08	30.02	196
130	0.188368	1043	0.0016	2662.6	3724.3	171.97	21.07	29.98	198
132	0.185364	1060	0.0016	2705.3	3784.3	172.42	21.06	29.94	199
134	0.182459	1078	0.0015	2748.0	3844.1	172.87	21.05	29.91	201
136	0.179647	1096	0.0015	2790.6	3903.9	173.32	21.05	29.89	202
138	0.176924	1114	0.0015	2833.2	3963.7	173.75	21.04	29.86	204
140	0.174285	1131	0.0015	2875.8	4023.4	174.18	21.04	29.84	205
142	0.171726	1149	0.0014	2918.4	4083.0	174.61	21.04	29.82	207
144	0.169245	1166	0.0014	2960.9	4142.6	175.02	21.04	29.80	209
146	0.166836	1184	0.0014	3003.4	4202.2	175.43	21.04	29.78	210
148	0.164497	1201	0.0014	3045.9	4261.8	175.84	21.04	29.77	211
150	0.162225	1219	0.0014	3088.4	4321.3	176.24	21.05	29.76	213
152	0.160016	1236	0.0013	3130.9	4380.8	176.63	21.05	29.75	214
154	0.157869	1253	0.0013	3173.4	4440.3	177.02	21.06	29.74	216
156	0.155780	1271	0.0013	3215.9	4499.8	177.40	21.06	29.73	217
158	0.153747	1288	0.0013	3258.4	4559.2	177.78	21.07	29.73	219
160	0.151768	1305	0.0013	3300.9	4618.7	178.16	21.08	29.73	220
165	0.147041	1349	0.0012	3407.1	4767.3	179.07	21.10	29.73	224
170	0.142605	1392	0.0012	3513.5	4915.9	179.96	21.13	29.73	227
175	0.138435	1434	0.0012	3619.9	5064.6	180.82	21.17	29.75	230
180	0.134506	1477	0.0011	3726.5	5213.4	181.66	21.21	29.77	234
185	0.130797	1520	0.0011	3833.3	5362.4	182.48	21.26	29.80	237
190	0.127291	1563	0.0011	3940.2	5511.4	183.27	21.31	29.84	240
195	0.123971	1605	0.0010	4047.4	5660.7	184.05	21.36	29.88	243
200	0.120821	1648	0.0010	4154.9	5810.2	184.80	21.42	29.92	246
205	0.117830	1690	0.0010	4262.6	5960.0	185.54	21.49	29.98	249
210	0.114985	1733	0.0010	4370.6	6110.0	186.27	21.55	30.03	252
215	0.112276	1775	0.0009	4479.0	6260.3	186.97	21.62	30.09	255
220	0.109693	1817	0.0009	4587.6	6410.9	187.67	21.69	30.16	258
225	0.107227	1860	0.0009	4696.7	6561.9	188.34	21.77	30.22	261
230	0.104870	1902	0.0009	4806.1	6713.2	189.01	21.85	30.29	263
235	0.102616	1944	0.0009	4915.8	6864.8	189.66	21.93	30.37	266
240	0.100458	1986	0.0008	5026.0	7016.8	190.30	22.01	30.44	269
245	0.0983894	2028	0.0008	5136.5	7169.2	190.93	22.09	30.52	272
250	0.0964048	2071	0.0008	5247.5	7322.1	191.55	22.18	30.60	274
255	0.0944994	2113	0.0008	5358.8	7475.3	192.15	22.27	30.68	277
260	0.0926684	2155	0.0008	5470.6	7628.9	192.75	22.35	30.77	279
265	0.0909074	2197	0.0008	5582.9	7782.9	193.34	22.44	30.85	282
270	0.0892126	2239	0.0007	5695.5	7937.4	193.92	22.53	30.93	284
275	0.0875803	2281	0.0007	5808.6	8092.3	194.48	22.62	31.02	287
280	0.0860069	2323	0.0007	5922.2	8247.6	195.04	22.71	31.11	289
285	0.0844895	2365	0.0007	6036.2	8403.3	195.59	22.81	31.20	292
290	0.0830250	2407	0.0007	6150.6	8559.5	196.14	22.90	31.28	294
295	0.0816107	2449	0.0007	6265.5	8716.2	196.67	22.99	31.37	297
300	0.0802440	2491	0.0007	6380.8	8873.2	197.20	23.08	31.46	299

* TWO-PHASE BOUNDARY

TABLE 16. THERMODYNAMIC PROPERTIES OF FLUORINE

0.3 MN/m² ISOBAR

TEMPERATURE (IPTS 1968) K	DENSITY MOL/L	ISOTHERM DERIVATIVE J/MOL	ISOCHORE DERIVATIVE MN/m²-K	INTERNAL ENERGY J/MOL	ENTHALPY J/MOL	ENTROPY J/MOL-K	C_v J / MOL - K	C_p J / MOL - K	VELOCITY OF SOUND M/S
* 53.510	44.8836	26865	4.3116	-6033.2	-6026.5	60.84	36.50	54.88	1031
54	44.8035	26820	4.3113	-6006.1	-5999.5	61.34	36.41	55.06	1033
56	44.4764	26361	4.3031	-5895.5	-5888.7	63.35	36.06	55.95	1037
58	44.1521	26002	4.2086	-5784.8	-5778.0	65.30	35.70	55.96	1036
60	43.8366	25787	4.1292	-5675.0	-5668.2	67.16	35.32	55.96	1037
- 62	43.5128	25548	4.0597	-5564.1	-5557.2	68.98	34.94	56.06	1039
64	43.1794	24800	3.9676	-5451.8	-5444.9	70.76	34.53	56.31	1032
66	42.8449	23656	3.8280	-5339.4	-5332.4	72.49	34.10	56.38	1014
68	42.5247	22819	3.6428	-5229.0	-5221.9	74.14	33.70	55.57	995
70	42.1949	21631	3.5211	-5117.5	-5110.3	75.76	33.29	55.83	977
72	41.8459	20425	3.4160	-5003.4	-4996.3	77.36	32.88	56.37	960
74	41.5066	20134	3.3358	-4890.9	-4883.7	78.90	32.49	56.23	958
76	41.1600	19194	3.1979	-4777.6	-4770.3	80.41	32.09	55.99	939
78	40.8095	17993	3.0807	-4664.0	-4656.6	81.89	31.71	56.41	918
80	40.4544	17119	2.9802	-4549.9	-4542.5	83.33	31.34	56.70	903
82	40.0849	16304	2.8653	-4434.0	-4426.5	84.76	30.96	56.66	886
84	39.7300	15465	2.7510	-4320.4	-4312.8	86.12	30.64	56.68	868
86	39.3601	14513	2.6309	-4204.9	-4197.3	87.47	30.32	56.79	846
88	38.9843	13655	2.5115	-4088.8	-4081.1	88.80	30.01	56.76	824
90	38.6024	12629	2.4431	-3972.1	-3964.3	90.11	29.75	58.30	807
92	38.2134	12027	2.3589	-3854.5	-3846.6	91.39	29.49	58.64	793
94	37.8170	11454	2.2641	-3736.0	-3728.1	92.65	29.25	58.67	778
* 95.782	37.4571	10686	2.1705	-3629.6	-3621.6	93.76	29.06	59.15	757
‡ 95.782	0.404713	686	0.0036	1879.8	2621.0	158.84	22.07	33.14	165
96	0.403571	688	0.0036	1884.9	2628.3	158.91	22.05	33.10	165
98	0.393451	710	0.0035	1931.6	2694.1	159.59	21.94	32.76	167
100	0.383921	731	0.0034	1977.9	2759.3	160.25	21.85	32.47	169
102	0.374922	752	0.0033	2023.9	2824.0	160.89	21.76	32.21	171
104	0.366403	773	0.0032	2069.4	2888.2	161.51	21.68	31.98	173
106	0.358320	793	0.0031	2114.7	2952.0	162.12	21.62	31.78	175
108	0.350636	813	0.0030	2159.7	3015.3	162.71	21.56	31.59	177
110	0.343317	833	0.0030	2204.5	3078.3	163.29	21.50	31.43	179
112	0.336335	853	0.0029	2249.1	3141.0	163.85	21.45	31.28	181
114	0.329664	872	0.0028	2293.4	3203.5	164.41	21.41	31.14	183
116	0.323281	892	0.0028	2337.6	3265.6	164.95	21.37	31.02	185
118	0.317165	911	0.0027	2381.7	3327.6	165.48	21.33	30.91	186
120	0.311298	930	0.0027	2425.6	3389.3	166.00	21.30	30.81	188
122	0.305664	949	0.0026	2469.3	3450.8	166.50	21.27	30.72	190
124	0.300248	968	0.0026	2513.0	3512.2	167.00	21.25	30.64	192
126	0.295035	986	0.0025	2556.5	3573.4	167.49	21.22	30.56	193
128	0.290015	1005	0.0025	2600.0	3634.4	167.97	21.20	30.49	195
130	0.285175	1023	0.0024	2643.3	3695.3	168.45	21.19	30.43	197
132	0.280504	1042	0.0024	2686.6	3756.1	168.91	21.17	30.37	198
134	0.275995	1060	0.0023	2729.8	3816.8	169.37	21.16	30.32	200
136	0.271637	1078	0.0023	2773.0	3877.4	169.81	21.15	30.27	202
138	0.267423	1096	0.0023	2816.0	3937.9	170.26	21.14	30.22	203
140	0.263345	1115	0.0022	2859.1	3998.3	170.69	21.13	30.18	205
142	0.259396	1133	0.0022	2902.1	4058.6	171.12	21.12	30.15	206
144	0.255570	1151	0.0022	2945.0	4118.9	171.54	21.12	30.11	208
146	0.251861	1169	0.0021	2987.9	4179.0	171.96	21.11	30.08	209
148	0.248263	1186	0.0021	3030.8	4239.2	172.36	21.11	30.06	211
150	0.244772	1204	0.0021	3073.6	4299.3	172.77	21.11	30.03	212
152	0.241381	1222	0.0020	3116.5	4359.3	173.17	21.11	30.01	214
154	0.238087	1240	0.0020	3159.3	4419.3	173.56	21.11	29.99	215
156	0.234886	1258	0.0020	3202.1	4479.3	173.94	21.12	29.98	217
158	0.231773	1275	0.0020	3244.9	4539.2	174.33	21.12	29.96	218
160	0.228744	1293	0.0019	3287.6	4599.1	174.70	21.13	29.95	220
165	0.221519	1337	0.0019	3394.5	4748.8	175.62	21.15	29.93	223
170	0.214752	1381	0.0018	3501.5	4898.4	176.52	21.17	29.92	227
175	0.208397	1424	0.0018	3608.5	5048.0	177.38	21.21	29.92	230
180	0.202418	1468	0.0017	3715.6	5197.6	178.23	21.24	29.93	233
185	0.196781	1511	0.0017	3822.8	5347.3	179.05	21.29	29.95	237
190	0.191457	1554	0.0016	3930.2	5497.1	179.85	21.33	29.97	240
195	0.186419	1597	0.0016	4037.8	5647.0	180.63	21.39	30.00	243
200	0.181645	1640	0.0015	4145.6	5797.1	181.39	21.44	30.04	246
205	0.177114	1683	0.0015	4253.6	5947.5	182.13	21.51	30.09	249
210	0.172808	1726	0.0014	4362.0	6098.0	182.85	21.57	30.14	252
215	0.168709	1769	0.0014	4470.6	6248.8	183.56	21.64	30.19	255
220	0.164804	1812	0.0014	4579.6	6399.9	184.26	21.71	30.25	258
225	0.161078	1854	0.0014	4688.8	6551.3	184.94	21.79	30.31	261
230	0.157519	1897	0.0013	4798.5	6703.0	185.61	21.86	30.38	263
235	0.154116	1939	0.0013	4908.5	6855.1	186.26	21.94	30.45	266
240	0.150859	1982	0.0013	5018.9	7007.5	186.90	22.02	30.52	269
245	0.147738	2024	0.0012	5129.6	7160.2	187.53	22.11	30.59	272
250	0.144746	2067	0.0012	5240.8	7313.4	188.15	22.19	30.67	274
255	0.141874	2109	0.0012	5352.4	7466.9	188.76	22.28	30.75	277
260	0.139114	2151	0.0012	5464.4	7620.9	189.36	22.37	30.83	279
265	0.136461	2194	0.0011	5576.8	7775.2	189.94	22.46	30.91	282
270	0.133909	2236	0.0011	5689.6	7930.0	190.52	22.55	30.99	284
275	0.131451	2278	0.0011	5802.9	8085.1	191.09	22.64	31.08	287
280	0.129082	2320	0.0011	5916.6	8240.7	191.65	22.73	31.16	289
285	0.126798	2362	0.0011	6030.8	8396.8	192.20	22.82	31.25	292
290	0.124595	2405	0.0010	6145.4	8553.2	192.75	22.91	31.33	294
295	0.122467	2447	0.0010	6260.5	8710.1	193.29	23.00	31.42	297
300	0.120411	2489	0.0010	6375.9	8867.4	193.81	23.09	31.51	299

* TWO-PHASE BOUNDARY

TABLE 16. THERMODYNAMIC PROPERTIES OF FLUORINE

0.4 MN/m² ISOBAR

TEMPERATURE (IPTS 1968) K	DENSITY MOL/L	ISOTHERM DERIVATIVE J/MOL	ISOCHORE DERIVATIVE MN/m²-K	INTERNAL ENERGY J/MOL	ENTHALPY J/MOL	ENTROPY J/MOL-K	C_v J / MOL - K	C_p J / MOL - K	VELOCITY OF SOUND M/S
* 53.520	44.8857	26907	4.3121	-6035.3	-6026.4	60.84	36.49	54.85	1032
54	44.8072	26859	4.3118	-6008.8	-5999.9	61.33	36.41	55.03	1034
56	44.4801	26387	4.3046	-5898.2	-5889.2	63.35	36.06	55.94	1038
58	44.1559	26020	4.2101	-5787.5	-5778.5	65.29	35.69	55.96	1036
60	43.8405	25804	4.1298	-5677.8	-5668.7	67.15	35.32	55.95	1037
62	43.5167	25544	4.0592	-5566.9	-5557.7	68.97	34.95	56.07	1039
64	43.1835	24778	3.9671	-5454.7	-5445.4	70.75	34.53	56.33	1031
66	42.8492	23642	3.8291	-5342.3	-5333.0	72.48	34.11	56.41	1014
68	42.5290	22824	3.6447	-5232.0	-5222.5	74.13	33.70	55.59	995
70	42.1995	21641	3.5221	-5120.5	-5111.0	75.75	33.30	55.83	977
72	41.8508	20423	3.4163	-5006.5	-4997.0	77.35	32.89	56.38	960
74	41.5116	20118	3.3355	-4894.1	-4884.4	78.89	32.50	56.25	957
76	41.1652	19186	3.1988	-4780.8	-4771.1	80.40	32.10	56.02	939
78	40.8151	17993	3.0811	-4667.2	-4657.4	81.88	31.72	56.42	918
80	40.4603	17120	2.9808	-4553.2	-4543.3	83.32	31.35	56.71	903
82	40.0911	16296	2.8661	-4437.3	-4427.4	84.75	30.97	56.69	886
84	39.7365	15472	2.7522	-4323.9	-4313.8	86.11	30.65	56.69	868
86	39.3670	14524	2.6324	-4208.4	-4198.3	87.46	30.32	56.80	846
88	38.9916	13665	2.5123	-4092.4	-4082.2	88.79	30.02	56.76	825
90	38.6103	12650	2.4432	-3975.8	-3965.5	90.09	29.77	58.26	807
92	38.2217	12043	2.3592	-3858.3	-3847.9	91.37	29.52	58.62	793
94	37.8257	11466	2.2647	-3739.9	-3729.4	92.64	29.27	58.66	778
96	37.4220	10639	2.1670	-3620.6	-3610.0	93.88	29.05	59.31	756
98	37.0092	10020	2.0803	-3500.2	-3489.4	95.11	28.85	59.76	739
* 99.167	36.7641	9545	2.0178	-3429.4	-3418.5	95.82	28.74	60.04	724
* 99.167	0.529979	685	0.0048	1920.9	2675.7	157.19	22.37	34.22	166
100	0.524224	694	0.0047	1941.1	2704.1	157.47	22.31	34.03	167
102	0.511035	717	0.0046	1989.0	2771.7	158.14	22.18	33.61	169
104	0.498641	740	0.0044	2036.4	2838.6	158.79	22.06	33.24	171
106	0.486959	762	0.0043	2083.3	2904.8	159.42	21.96	32.92	173
108	0.475917	783	0.0042	2129.8	2970.3	160.04	21.87	32.64	175
110	0.465454	804	0.0041	2176.0	3035.3	160.63	21.78	32.39	177
112	0.455519	825	0.0040	2221.8	3099.9	161.21	21.71	32.16	179
114	0.446065	846	0.0039	2267.3	3164.0	161.78	21.64	31.96	181
116	0.437053	866	0.0038	2312.5	3227.7	162.33	21.58	31.78	183
118	0.428447	886	0.0037	2357.5	3291.1	162.88	21.53	31.61	185
120	0.420219	906	0.0036	2402.3	3354.2	163.41	21.48	31.46	187
122	0.412338	926	0.0036	2446.9	3417.0	163.93	21.44	31.33	189
124	0.404782	946	0.0035	2491.3	3479.5	164.43	21.40	31.21	190
126	0.397528	965	0.0034	2535.6	3541.8	164.93	21.37	31.09	192
128	0.390556	984	0.0034	2579.7	3603.9	165.42	21.34	30.99	194
130	0.383848	1004	0.0033	2623.7	3665.8	165.90	21.31	30.90	196
132	0.377389	1023	0.0032	2667.6	3727.5	166.37	21.29	30.81	197
134	0.371162	1042	0.0032	2711.3	3789.0	166.83	21.27	30.74	199
136	0.365154	1060	0.0031	2755.0	3850.4	167.29	21.25	30.67	201
138	0.359354	1079	0.0031	2798.6	3911.7	167.74	21.23	30.60	202
140	0.353749	1098	0.0030	2842.1	3972.9	168.18	21.22	30.54	204
142	0.348329	1116	0.0030	2885.5	4033.9	168.61	21.21	30.49	206
144	0.343083	1135	0.0029	2928.9	4094.8	169.04	21.20	30.44	207
146	0.338005	1153	0.0029	2972.2	4155.6	169.46	21.19	30.39	209
148	0.333084	1172	0.0028	3015.5	4216.4	169.87	21.18	30.35	210
150	0.328312	1190	0.0028	3058.7	4277.0	170.28	21.18	30.32	212
152	0.323684	1208	0.0028	3101.9	4337.6	170.68	21.17	30.28	213
154	0.319192	1226	0.0027	3145.0	4398.2	171.07	21.17	30.25	215
156	0.314829	1244	0.0027	3188.1	4458.7	171.46	21.17	30.22	216
158	0.310591	1262	0.0026	3231.2	4519.1	171.85	21.17	30.20	218
160	0.306470	1280	0.0026	3274.3	4579.5	172.23	21.18	30.18	219
165	0.296654	1325	0.0025	3381.9	4730.2	173.16	21.19	30.14	223
170	0.287474	1370	0.0024	3489.4	4880.8	174.05	21.21	30.11	226
175	0.278866	1414	0.0024	3596.9	5031.3	174.93	21.24	30.09	230
180	0.270778	1458	0.0023	3704.5	5181.8	175.77	21.27	30.09	233
185	0.263161	1502	0.0022	3812.2	5332.2	176.60	21.31	30.09	236
190	0.255974	1546	0.0022	3920.0	5482.7	177.40	21.36	30.11	239
195	0.249180	1589	0.0021	4028.0	5633.3	178.18	21.41	30.13	243
200	0.242747	1633	0.0020	4136.2	5784.0	178.95	21.47	30.16	246
205	0.236646	1676	0.0020	4244.6	5934.9	179.69	21.53	30.20	249
210	0.230852	1720	0.0019	4353.3	6086.0	180.42	21.59	30.24	252
215	0.225341	1763	0.0019	4462.2	6237.3	181.13	21.66	30.29	255
220	0.220092	1806	0.0019	4571.5	6388.9	181.83	21.73	30.34	258
225	0.215087	1849	0.0018	4681.0	6540.7	182.51	21.80	30.40	260
230	0.210310	1892	0.0018	4790.9	6692.9	183.18	21.88	30.46	263
235	0.205743	1935	0.0017	4901.1	6845.3	183.84	21.96	30.52	266
240	0.201374	1977	0.0017	5011.8	6998.1	184.48	22.04	30.59	269
245	0.197190	2020	0.0017	5122.7	7151.2	185.11	22.12	30.66	271
250	0.193179	2063	0.0016	5234.1	7304.7	185.73	22.21	30.74	274
255	0.189330	2105	0.0016	5345.9	7458.6	186.34	22.29	30.81	277
260	0.185634	2148	0.0016	5458.1	7612.9	186.94	22.38	30.89	279
265	0.182082	2190	0.0015	5570.7	7767.5	187.53	22.47	30.97	282
270	0.178664	2233	0.0015	5683.7	7922.6	188.11	22.56	31.05	284
275	0.175375	2275	0.0015	5797.2	8078.0	188.68	22.65	31.13	287
280	0.172205	2318	0.0014	5911.1	8233.9	189.24	22.74	31.22	289
285	0.169150	2360	0.0014	6025.4	8390.2	189.79	22.83	31.30	292
290	0.166202	2402	0.0014	6140.2	8546.9	190.34	22.92	31.38	294
295	0.163357	2445	0.0014	6255.4	8704.0	190.88	23.02	31.47	297
300	0.160608	2487	0.0013	6371.0	8861.6	191.41	23.11	31.55	299

* TWO-PHASE BOUNDARY

TABLE 16. THERMODYNAMIC PROPERTIES OF FLUORINE

MN/m² ISOBAR

DENSITY MOL/L	ISOTHERM DERIVATIVE J/MOL	ISOCHORE DERIVATIVE MN/m²-K	INTERNAL ENERGY J/MOL	ENTHALPY J/MOL	ENTROPY J/MOL-K	C_v J / MOL - K	C_p J / MOL - K	VELOCITY OF SOUND M/S
44.8878	26948	4.3126	-6037.4	-6026.2	60.84	36.49	54.82	1032
44.8109	26898	4.3122	-6011.5	-6000.3	61.32	36.41	55.00	1034
44.6839	26413	4.3061	-5900.9	-5889.7	63.34	36.06	55.93	1038
44.1598	26038	4.2116	-5790.3	-5779.0	65.28	35.69	55.95	1036
43.8444	25821	4.1304	-5680.6	-5669.2	67.14	35.32	55.95	1037
43.5207	25540	4.0587	-5569.7	-5558.2	68.96	34.95	56.07	1038
43.1875	24757	3.9667	-5457.5	-5446.0	70.75	34.54	56.35	1031
42.8534	23628	3.8301	-5345.2	-5333.6	72.48	34.12	56.43	1014
42.5334	22828	3.6466	-5234.9	-5223.1	74.12	33.71	55.60	996
42.2041	21651	3.5231	-5123.5	-5111.6	75.74	33.30	55.83	977
41.8597	20421	3.4165	-5009.6	-4997.6	77.34	32.89	56.39	960
41.5166	20104	3.3353	-4897.2	-4885.1	78.88	32.51	56.27	957
41.1705	19178	3.1998	-4783.9	-4771.8	80.39	32.11	56.04	939
40.8207	17994	3.0816	-4670.5	-4658.2	81.87	31.72	56.43	918
40.4661	17122	2.9813	-4556.5	-4544.2	83.31	31.36	56.72	903
40.0972	16288	2.8669	-4440.7	-4428.3	84.74	30.98	56.72	886
39.7429	15479	2.7533	-4327.3	-4314.7	86.10	30.66	56.70	868
39.3738	14534	2.6338	-4212.0	-4199.3	87.45	30.33	56.81	846
38.9989	13676	2.5132	-4096.0	-4083.2	88.78	30.03	56.75	825
38.6182	12672	2.4433	-3979.6	-3966.6	90.08	29.79	58.22	807
38.2300	12058	2.3595	-3862.2	-3849.1	91.36	29.54	58.60	793
37.8344	11477	2.2653	-3743.9	-3730.7	92.62	29.29	58.65	778
37.4314	10657	2.1677	-3624.7	-3611.3	93.87	29.07	59.28	756
37.0192	10037	2.0808	-3504.4	-3490.9	95.09	28.88	59.73	739
36.5975	9311	1.9915	-3382.8	-3369.1	96.30	28.69	60.49	719
36.1717	8759	1.8952	-3261.7	-3247.9	97.49	28.53	60.48	699
0.654419	680	0.0060	1951.1	2715.1	155.89	22.66	35.28	167
0.654146	680	0.0060	1951.9	2716.2	155.91	22.66	35.27	167
0.637109	705	0.0058	2001.4	2786.2	156.58	22.49	34.72	169
0.621175	729	0.0056	2050.2	2855.2	157.24	22.34	34.25	171
0.606217	752	0.0055	2098.4	2923.2	157.88	22.21	33.83	174
0.592129	775	0.0053	2146.1	2990.5	158.50	22.09	33.47	176
0.578821	797	0.0052	2193.3	3057.1	159.10	21.99	33.15	178
0.566218	819	0.0050	2240.1	3123.1	159.68	21.90	32.86	180
0.554255	840	0.0049	2286.5	3188.6	160.25	21.82	32.61	182
0.542876	861	0.0048	2332.6	3253.6	160.80	21.75	32.38	184
0.532031	882	0.0047	2378.3	3318.1	161.35	21.68	32.17	186
0.521679	903	0.0046	2423.8	3382.3	161.88	21.62	31.99	188
0.511780	923	0.0045	2469.1	3446.1	162.40	21.57	31.82	189
0.502302	944	0.0044	2514.1	3509.6	162.90	21.52	31.67	191
0.493214	964	0.0043	2559.0	3572.8	163.40	21.48	31.53	193
0.484490	984	0.0042	2603.7	3635.7	163.89	21.44	31.40	195
0.476105	1003	0.0041	2648.2	3698.4	164.37	21.41	31.29	196
0.468037	1023	0.0040	2692.5	3760.8	164.84	21.38	31.18	198
0.460267	1042	0.0040	2736.8	3823.1	165.30	21.35	31.09	200
0.452777	1062	0.0039	2780.9	3885.2	165.75	21.33	31.00	202
0.445550	1081	0.0038	2824.9	3947.1	166.20	21.31	30.92	203
0.438571	1100	0.0038	2868.8	4008.9	166.63	21.29	30.84	205
0.431827	1119	0.0037	2912.6	4070.5	167.07	21.28	30.78	206
0.425304	1138	0.0037	2956.3	4132.0	167.49	21.26	30.71	208
0.418991	1157	0.0036	3000.0	4193.3	167.91	21.25	30.66	210
0.412877	1175	0.0035	3043.6	4254.6	168.32	21.24	30.61	211
0.406952	1194	0.0035	3087.1	4315.8	168.72	21.24	30.56	213
0.401206	1212	0.0034	3130.6	4376.8	169.12	21.23	30.52	214
0.395632	1231	0.0034	3174.0	4437.8	169.52	21.23	30.48	216
0.390221	1249	0.0033	3217.4	4498.8	169.90	21.23	30.44	217
0.384965	1268	0.0033	3260.8	4559.6	170.29	21.23	30.41	219
0.372459	1313	0.0032	3369.1	4711.5	171.22	21.24	30.35	222
0.360782	1359	0.0031	3477.2	4863.1	172.13	21.25	30.30	226
0.349850	1404	0.0030	3585.4	5014.5	173.01	21.27	30.27	229
0.339591	1448	0.0029	3693.5	5165.8	173.86	21.31	30.25	233
0.329941	1493	0.0028	3801.6	5317.0	174.69	21.34	30.24	236
0.320845	1537	0.0027	3909.9	5468.3	175.49	21.39	30.25	239
0.312255	1581	0.0026	4018.3	5619.5	176.28	21.43	30.26	242
0.304129	1625	0.0026	4126.8	5770.9	177.04	21.49	30.28	246
0.296428	1669	0.0025	4235.6	5922.3	177.79	21.55	30.31	249
0.289119	1713	0.0024	4344.6	6074.0	178.52	21.61	30.34	252
0.282171	1757	0.0024	4453.8	6225.8	179.24	21.68	30.39	255
0.275558	1800	0.0023	4563.3	6377.8	179.94	21.75	30.43	257
0.269256	1843	0.0023	4673.2	6530.1	180.62	21.82	30.49	260
0.263243	1887	0.0022	4783.3	6682.7	181.29	21.89	30.54	263
0.257498	1930	0.0022	4893.8	6835.5	181.95	21.97	30.60	266
0.252004	1973	0.0021	5004.6	6988.7	182.59	22.05	30.67	269
0.246745	2016	0.0021	5115.8	7142.2	183.23	22.14	30.74	271
0.241704	2059	0.0020	5227.4	7296.1	183.85	22.22	30.81	274
0.236870	2102	0.0020	5339.4	7450.3	184.46	22.31	30.88	277
0.232228	2145	0.0020	5451.8	7604.9	185.06	22.40	30.95	279
0.227768	2187	0.0019	5564.6	7759.8	185.65	22.48	31.03	282
0.223479	2230	0.0019	5677.8	7915.2	186.23	22.57	31.11	284
0.219351	2273	0.0018	5791.5	8070.9	186.80	22.66	31.19	287
0.215375	2315	0.0018	5905.5	8227.1	187.37	22.76	31.27	289
0.211543	2358	0.0018	6020.0	8383.6	187.92	22.85	31.35	292
0.207847	2400	0.0017	6135.0	8540.6	188.47	22.94	31.43	294
0.204280	2443	0.0017	6250.3	8698.0	189.00	23.03	31.52	297
0.200834	2485	0.0017	6366.1	8855.7	189.53	23.12	31.60	299

SE BOUNDARY

TABLE 16. THERMODYNAMIC PROPERTIES OF FLUORINE

0.6 MN/m² ISOBAR

TEMPERATURE (IPTS 1968) K	DENSITY MOL/L	ISOTHERM DERIVATIVE J/MOL	ISOCHORE DERIVATIVE MN/m²·K	INTERNAL ENERGY J/MOL	ENTHALPY J/MOL	ENTROPY J/MOL-K	C_v J / MOL - K	C_p J / MOL - K	VELOCITY OF SOUND M/S
* 53.540	44.8899	26990	4.3130	-6039.5	-6026.1	60.84	36.48	54.80	1033
54	44.8147	26938	4.3127	-6014.1	-6000.7	61.32	36.41	54.97	1035
56	44.4877	26440	4.3076	-5903.6	-5890.1	63.33	36.06	55.92	1039
58	44.1636	26056	4.2131	-5793.0	-5779.4	65.27	35.69	55.95	1037
60	43.8482	25838	4.1311	-5683.4	-5669.7	67.13	35.32	55.94	1038
62	43.5246	25537	4.0582	-5572.5	-5558.7	68.95	34.96	56.07	1038
64	43.1915	24736	3.9662	-5460.4	-5446.5	70.74	34.55	56.37	1031
66	42.8576	23614	3.8312	-5348.1	-5334.1	72.47	34.13	56.46	1014
68	42.5378	22832	3.6485	-5237.8	-5223.7	74.11	33.71	55.62	996
70	42.2087	21661	3.5241	-5126.5	-5112.3	75.73	33.31	55.83	978
72	41.8606	20419	3.4168	-5012.7	-4998.3	77.33	32.90	56.39	960
74	41.5216	20089	3.3350	-4900.3	-4885.9	78.88	32.52	56.29	957
76	41.1757	19171	3.2007	-4787.1	-4772.5	80.38	32.12	56.07	939
78	40.8262	17995	3.0820	-4673.7	-4659.0	81.86	31.73	56.43	918
80	40.4720	17124	2.9819	-4559.8	-4545.0	83.30	31.37	56.73	903
82	40.1034	16281	2.8676	-4444.1	-4429.2	84.72	30.99	56.75	886
84	39.7494	15486	2.7544	-4330.8	-4315.7	86.09	30.67	56.71	868
86	39.3807	14544	2.6393	-4215.5	-4200.3	87.44	30.34	56.82	847
88	39.0062	13687	2.5149	-4099.7	-4084.3	88.77	30.03	56.76	825
90	38.6261	12693	2.4435	-3983.3	-3967.8	90.07	29.80	58.18	808
92	38.2383	12074	2.3598	-3866.0	-3850.3	91.35	29.56	58.58	794
94	37.8431	11489	2.2660	-3747.8	-3732.0	92.61	29.31	58.64	778
96	37.4408	10675	2.1684	-3628.8	-3612.7	93.85	29.09	59.25	756
98	37.0292	10054	2.0814	-3508.5	-3492.3	95.08	28.90	59.70	739
100	36.6082	9331	1.9924	-3387.1	-3370.7	96.29	28.71	60.46	719
102	36.1763	8768	1.9080	-3264.3	-3247.7	97.49	28.55	60.91	702
104	35.7328	8208	1.8169	-3139.8	-3123.0	98.68	28.39	61.15	682
* 104.383	35.6463	8108	1.7924	-3115.8	-3098.9	98.90	28.37	60.92	677
* 104.383	0.778634	673	0.0072	1973.9	2744.5	154.82	22.93	36.33	168
106	0.761762	694	0.0070	2015.2	2802.8	155.38	22.77	35.79	169
108	0.742184	719	0.0068	2065.4	2873.8	156.04	22.59	35.21	172
110	0.723874	744	0.0066	2114.8	2943.7	156.68	22.44	34.70	174
112	0.706686	767	0.0064	2163.6	3012.6	157.30	22.30	34.26	176
114	0.690496	791	0.0062	2211.8	3080.8	157.91	22.18	33.87	178
116	0.675201	813	0.0061	2259.5	3148.1	158.49	22.07	33.52	180
118	0.660715	836	0.0059	2306.8	3214.9	159.06	21.98	33.22	182
120	0.646963	858	0.0058	2353.6	3281.0	159.62	21.89	32.94	184
122	0.633879	880	0.0056	2400.1	3346.7	160.16	21.81	32.70	186
124	0.621408	901	0.0055	2446.3	3411.8	160.69	21.75	32.48	188
126	0.609501	922	0.0054	2492.2	3476.6	161.21	21.68	32.28	190
128	0.598113	943	0.0053	2537.8	3541.0	161.72	21.63	32.10	192
130	0.587207	963	0.0051	2583.2	3605.0	162.21	21.58	31.93	194
132	0.576747	984	0.0050	2628.4	3668.7	162.70	21.54	31.78	195
134	0.566703	1004	0.0049	2673.4	3732.1	163.18	21.50	31.65	197
136	0.557048	1024	0.0048	2718.2	3795.3	163.64	21.46	31.53	199
138	0.547756	1044	0.0048	2762.9	3858.3	164.10	21.43	31.41	201
140	0.538805	1064	0.0047	2807.4	3921.0	164.55	21.40	31.31	202
142	0.530174	1083	0.0046	2851.8	3983.5	165.00	21.38	31.21	204
144	0.521844	1103	0.0045	2896.1	4045.8	165.43	21.36	31.13	206
146	0.513798	1122	0.0044	2940.2	4108.0	165.86	21.34	31.05	207
148	0.506020	1142	0.0044	2984.3	4170.0	166.28	21.33	30.98	209
150	0.498496	1161	0.0043	3028.3	4231.9	166.70	21.31	30.91	210
152	0.491212	1180	0.0042	3072.2	4293.7	167.11	21.30	30.85	212
154	0.484156	1199	0.0042	3116.0	4355.3	167.51	21.29	30.79	214
156	0.477317	1218	0.0041	3159.8	4416.8	167.91	21.29	30.74	215
158	0.470682	1236	0.0040	3203.5	4478.3	168.30	21.28	30.70	217
160	0.464244	1255	0.0040	3247.2	4539.6	168.69	21.28	30.65	218
165	0.448946	1302	0.0038	3356.2	4692.7	169.63	21.28	30.56	222
170	0.434686	1348	0.0037	3465.0	4845.3	170.54	21.29	30.50	225
175	0.421357	1393	0.0036	3573.7	4997.7	171.42	21.31	30.45	229
180	0.408864	1439	0.0035	3682.3	5149.8	172.28	21.34	30.41	232
185	0.397127	1484	0.0034	3791.0	5301.8	173.11	21.37	30.39	236
190	0.386076	1529	0.0033	3899.7	5453.8	173.92	21.41	30.39	239
195	0.375649	1573	0.0032	4008.5	5605.7	174.71	21.46	30.39	242
200	0.365793	1618	0.0031	4117.4	5757.7	175.48	21.51	30.40	245
205	0.356461	1662	0.0030	4226.5	5909.7	176.23	21.57	30.42	248
210	0.347609	1706	0.0029	4335.8	6061.9	176.97	21.63	30.45	251
215	0.339201	1750	0.0029	4445.4	6214.2	177.68	21.69	30.49	254
220	0.331203	1794	0.0028	4555.2	6366.8	178.38	21.76	30.53	257
225	0.323585	1838	0.0027	4665.3	6519.5	179.07	21.83	30.57	260
230	0.316319	1882	0.0027	4775.7	6672.5	179.74	21.91	30.63	263
235	0.309381	1925	0.0026	4886.4	6825.8	180.40	21.99	30.68	266
240	0.302749	1968	0.0026	4997.5	6979.3	181.05	22.07	30.74	269
245	0.296402	2012	0.0025	5108.9	7133.2	181.68	22.15	30.81	271
250	0.290321	2055	0.0024	5220.7	7287.4	182.31	22.24	30.87	274
255	0.284491	2098	0.0024	5332.9	7442.0	182.92	22.32	30.94	277
260	0.278896	2141	0.0023	5445.5	7596.9	183.52	22.41	31.02	279
265	0.273521	2184	0.0023	5558.5	7752.1	184.11	22.50	31.09	282
270	0.268353	2227	0.0023	5671.9	7907.8	184.69	22.59	31.17	284
275	0.263380	2270	0.0022	5785.7	8063.8	185.27	22.68	31.24	287
280	0.258592	2313	0.0022	5900.0	8220.2	185.83	22.77	31.32	289
285	0.253978	2355	0.0021	6014.6	8377.0	186.39	22.86	31.40	292
290	0.249529	2398	0.0021	6129.7	8534.3	186.93	22.95	31.48	294
295	0.245236	2441	0.0021	6245.3	8691.9	187.47	23.04	31.57	297
300	0.241090	2483	0.0020	6361.2	8849.9	188.00	23.14	31.65	299

* TWO-PHASE BOUNDARY

TABLE 16. THERMODYNAMIC PROPERTIES OF FLUORINE

.7 MN/m² ISOBAR

URE 68)	DENSITY MOL/L	ISOTHERM DERIVATIVE J/MOL	ISOCHORE DERIVATIVE MN/m²-K	INTERNAL ENERGY J/MOL	ENTHALPY J/MOL	ENTROPY J/MOL-K	C_v J / MOL - K	C_p J / MOL - K	VELOCITY OF SOUND M/S
550	44.8920	27031	4.3135	-6041.6	-6026.0	60.85	36.48	54.77	1033
	44.8184	26977	4.3132	-6016.8	-6001.2	61.31	36.41	54.94	1035
	44.4915	26466	4.3091	-5906.3	-5890.6	63.32	36.06	55.91	1039
	44.1674	26074	4.2145	-5795.8	-5779.9	65.26	35.69	55.94	1037
	43.8521	25856	4.1318	-5686.2	-5670.2	67.13	35.33	55.93	1038
	43.5285	25533	4.0578	-5575.3	-5559.3	68.95	34.97	56.07	1038
	43.1956	24716	3.9658	-5463.3	-5447.1	70.73	34.56	56.39	1030
	42.8619	23601	3.8322	-5351.0	-5334.7	72.46	34.14	56.49	1014
	42.5422	22837	3.6504	-5240.8	-5224.3	74.11	33.72	55.64	996
	42.2133	21671	3.5251	-5129.5	-5112.9	75.72	33.31	55.84	978
	41.8655	20418	3.4171	-5015.7	-4999.0	77.33	32.91	56.40	960
	41.5265	20075	3.3348	-4903.4	-4886.6	78.87	32.53	56.31	956
	41.1809	19164	3.2016	-4790.3	-4773.3	80.37	32.12	56.09	938
	40.8318	17997	3.0825	-4676.9	-4659.8	81.85	31.74	56.44	918
	40.4778	17126	2.9824	-4563.2	-4545.9	83.29	31.38	56.74	903
	40.1095	16273	2.8664	-4447.5	-4430.1	84.71	31.00	56.77	886
	39.7559	15493	2.7555	-4334.2	-4316.6	86.08	30.68	56.72	868
	39.3876	14555	2.6367	-4219.1	-4201.3	87.43	30.35	56.82	847
	39.0136	13697	2.5168	-4103.3	-4085.3	88.75	30.04	56.77	825
	38.6340	12714	2.4438	-3987.1	-3969.0	90.05	29.82	58.14	808
	38.2466	12090	2.3601	-3869.9	-3851.6	91.33	29.58	58.55	794
	37.8518	11501	2.2666	-3751.7	-3733.2	92.60	29.33	58.64	778
	37.4501	10693	2.1691	-3632.8	-3614.1	93.84	29.11	59.23	757
	37.0391	10072	2.0820	-3512.7	-3493.8	95.06	28.92	59.67	739
	36.6189	9351	1.9934	-3391.4	-3372.3	96.27	28.73	60.42	719
	36.1877	8786	1.9089	-3268.7	-3249.4	97.47	28.57	60.87	702
	35.7449	8224	1.8183	-3144.4	-3124.8	98.66	28.41	61.14	682
	35.2892	7492	1.7245	-3018.4	-2998.5	99.84	28.28	62.06	658
518	35.1688	7265	1.7038	-2985.4	-2965.5	100.14	28.24	62.65	651
518	0.903026	665	0.0085	1991.6	2766.7	153.90	23.19	37.39	168
	0.884616	685	0.0083	2030.4	2821.7	154.42	23.02	36.80	170
	0.861338	711	0.0080	2081.9	2894.6	155.09	22.82	36.11	172
	0.839642	737	0.0077	2132.5	2966.2	155.73	22.64	35.52	174
	0.819334	762	0.0075	2182.4	3036.7	156.35	22.49	35.00	177
	0.800254	786	0.0073	2231.5	3106.2	156.96	22.35	34.54	179
	0.782270	810	0.0071	2280.1	3174.9	157.55	22.22	34.14	181
	0.765269	833	0.0069	2328.1	3242.8	158.12	22.12	33.79	183
	0.749157	855	0.0067	2375.7	3310.1	158.67	22.02	33.47	185
	0.733853	878	0.0066	2422.9	3376.7	159.21	21.93	33.19	187
	0.719284	900	0.0064	2469.7	3442.9	159.74	21.86	32.94	189
	0.705391	922	0.0063	2516.1	3508.5	160.26	21.79	32.71	191
	0.692118	943	0.0061	2562.3	3573.7	160.77	21.72	32.50	193
	0.679419	964	0.0060	2608.2	3638.5	161.26	21.67	32.31	195
	0.667251	985	0.0059	2653.9	3703.0	161.75	21.62	32.14	196
	0.655576	1006	0.0058	2699.3	3767.1	162.22	21.58	31.99	198
	0.644361	1026	0.0056	2744.6	3830.9	162.69	21.54	31.85	200
	0.633575	1047	0.0055	2789.6	3894.5	163.14	21.50	31.72	202
	0.623191	1067	0.0054	2834.6	3957.8	163.59	21.47	31.60	203
	0.613183	1087	0.0053	2879.3	4020.9	164.03	21.44	31.49	205
	0.603529	1107	0.0052	2923.9	4083.8	164.47	21.42	31.39	207
	0.594209	1126	0.0052	2968.4	4146.5	164.89	21.40	31.30	208
	0.585204	1146	0.0051	3012.8	4209.0	165.31	21.38	31.22	210
	0.576495	1165	0.0050	3057.1	4271.4	165.73	21.37	31.14	211
	0.568067	1185	0.0049	3101.3	4333.6	166.13	21.36	31.08	213
	0.559906	1204	0.0048	3145.5	4395.7	166.53	21.35	31.01	215
	0.551997	1223	0.0048	3189.5	4457.6	166.93	21.34	30.95	216
	0.544328	1242	0.0047	3233.5	4519.5	167.32	21.33	30.90	218
	0.526130	1290	0.0045	3343.2	4673.7	168.27	21.33	30.79	221
	0.509198	1337	0.0044	3452.7	4827.4	169.18	21.33	30.70	225
	0.493395	1383	0.0042	3562.0	4980.7	170.07	21.35	30.63	229
	0.478603	1429	0.0041	3671.1	5133.7	170.94	21.37	30.58	232
	0.464723	1475	0.0040	3780.3	5286.5	171.77	21.40	30.55	235
	0.451668	1520	0.0039	3889.4	5439.2	172.59	21.44	30.53	239
	0.439363	1566	0.0037	3998.6	5591.8	173.38	21.48	30.52	242
	0.427742	1611	0.0036	4107.9	5744.4	174.15	21.53	30.52	245
	0.416746	1655	0.0035	4217.4	5897.1	174.91	21.59	30.54	248
	0.406325	1700	0.0035	4327.0	6049.8	175.64	21.65	30.56	251
	0.396432	1744	0.0034	4436.9	6202.7	176.36	21.71	30.59	254
	0.387027	1788	0.0033	4547.0	6355.7	177.07	21.78	30.62	257
	0.378073	1833	0.0032	4657.4	6508.9	177.75	21.85	30.66	260
	0.369538	1877	0.0031	4768.1	6662.3	178.43	21.93	30.71	263
	0.361392	1920	0.0031	4879.0	6816.0	179.09	22.00	30.76	266
	0.353607	1964	0.0030	4990.4	6970.0	179.74	22.08	30.82	269
	0.346161	2008	0.0029	5102.0	7124.2	180.37	22.17	30.88	271
	0.339030	2051	0.0029	5214.0	7278.8	181.00	22.25	30.94	274
	0.332195	2094	0.0028	5326.4	7433.6	181.61	22.34	31.01	277
	0.325637	2138	0.0027	5439.2	7588.9	182.21	22.42	31.08	279
	0.319339	2181	0.0027	5552.4	7744.4	182.81	22.51	31.15	282
	0.313285	2224	0.0026	5666.0	7900.4	183.39	22.60	31.23	284
	0.307462	2267	0.0026	5780.0	8056.7	183.96	22.69	31.30	287
	0.301856	2310	0.0025	5894.4	8213.4	184.53	22.78	31.38	289
	0.296455	2353	0.0025	6009.2	8370.5	185.08	22.87	31.46	292
	0.291248	2396	0.0025	6124.5	8527.9	185.63	22.97	31.53	294
	0.286224	2439	0.0024	6240.2	8685.8	186.17	23.06	31.61	297
	0.281374	2481	0.0024	6356.3	8844.1	186.70	23.15	31.69	299

=PHASE BOUNDARY

TABLE 16. THERMODYNAMIC PROPERTIES OF FLUORINE

0.8 MN/m² ISOBAR

TEMPERATURE (IPTS 1968) K	DENSITY MOL/L	ISOTHERM DERIVATIVE J/MOL	ISOCHORE DERIVATIVE MN/m²-K	INTERNAL ENERGY J/MOL	ENTHALPY J/MOL	ENTROPY J/MOL-K	C_V J / MOL - K	C_p	VELOCITY OF SOUND M/S
* 53.559	44.8941	27072	4.3140	-6043.7	-6025.9	60.85	36.48	54.75	1034
* 54	44.8221	27016	4.3136	-6019.4	-6001.6	61.30	36.40	54.92	1036
56	44.4953	26493	4.3106	-5909.0	-5891.0	63.31	36.06	55.89	1040
58	44.1713	26091	4.2160	-5798.5	-5780.4	65.26	35.69	55.94	1037
60	43.8560	25871	4.1326	-5688.9	-5670.7	67.12	35.33	55.92	1038
62	43.5324	25530	4.0574	-5578.2	-5559.8	68.94	34.98	56.07	1038
64	43.1996	24696	3.9653	-5466.1	-5447.6	70.72	34.57	56.40	1030
66	42.8661	23587	3.8333	-5353.9	-5335.3	72.45	34.14	56.52	1014
68	42.5466	22841	3.6522	-5243.7	-5224.9	74.10	33.72	55.66	996
70	42.2179	21681	3.5261	-5132.5	-5113.5	75.71	33.31	55.84	978
72	41.8704	20416	3.4174	-5018.8	-4999.7	77.32	32.91	56.41	960
74	41.5315	20061	3.3346	-4906.5	-4887.3	78.86	32.54	56.32	956
76	41.1861	19157	3.2025	-4793.5	-4774.0	80.37	32.13	56.12	938
78	40.8373	17998	3.0829	-4680.2	-4660.6	81.84	31.75	56.45	918
80	40.4836	17128	2.9830	-4566.5	-4546.7	83.28	31.39	56.75	903
82	40.1156	16266	2.8692	-4450.9	-4431.0	84.70	31.01	56.80	885
84	39.7623	15500	2.7566	-4337.7	-4317.6	86.06	30.68	56.73	868
86	39.3945	14566	2.6381	-4222.6	-4202.3	87.42	30.35	56.83	847
88	39.0208	13709	2.5186	-4106.9	-4086.4	88.74	30.04	56.79	826
90	38.6418	12735	2.4440	-3990.8	-3970.1	90.04	29.84	58.11	808
92	38.2548	12106	2.3604	-3873.7	-3852.8	91.32	29.60	58.53	794
94	37.8605	11513	2.2672	-3755.7	-3734.5	92.58	29.35	58.63	778
96	37.4595	10712	2.1699	-3636.9	-3615.5	93.82	29.13	59.20	757
98	37.0490	10089	2.0826	-3516.9	-3495.3	95.05	28.95	59.64	740
100	36.6296	9372	1.9943	-3395.2	-3373.3	96.26	28.75	60.38	720
102	36.1991	8604	1.9097	-3273.1	-3251.0	97.45	28.60	60.84	702
104	35.7571	8240	1.8197	-3149.0	-3126.6	98.64	28.43	61.12	683
106	35.3026	7515	1.7263	-3023.2	-3000.5	99.82	28.29	62.02	658
108	34.8330	6869	1.6421	-2895.3	-2872.3	100.99	28.17	63.12	636
* 108.440	34.7275	6740	1.6257	-2866.9	-2843.8	101.25	28.15	63.41	632
* 108.440	1.02789	656	0.0098	2005.2	2783.5	153.09	23.45	38.46	168
110	1.00532	678	0.0095	2047.1	2842.9	153.64	23.25	37.74	170
112	0.978328	705	0.0092	2099.8	2917.6	154.31	23.02	36.95	173
114	0.953250	732	0.0089	2151.5	2990.8	154.96	22.82	36.27	175
116	0.929838	758	0.0086	2202.3	3062.7	155.59	22.65	35.68	177
118	0.907891	783	0.0083	2252.4	3133.5	156.19	22.49	35.17	179
120	0.887245	807	0.0081	2301.7	3203.4	156.78	22.36	34.71	182
122	0.867760	831	0.0079	2350.5	3272.4	157.35	22.24	34.31	184
124	0.849322	854	0.0077	2398.8	3340.7	157.90	22.13	33.96	186
126	0.831830	877	0.0075	2446.6	3408.3	158.44	22.04	33.64	188
128	0.815199	900	0.0073	2493.9	3475.3	158.97	21.95	33.36	190
130	0.799355	922	0.0071	2540.9	3541.7	159.49	21.88	33.10	192
132	0.784233	944	0.0070	2587.6	3607.7	159.99	21.81	32.87	194
134	0.769777	966	0.0068	2634.0	3673.3	160.48	21.75	32.67	195
136	0.755936	987	0.0067	2680.1	3738.4	160.97	21.69	32.48	197
138	0.742665	1008	0.0066	2726.0	3803.2	161.44	21.65	32.30	199
140	0.729924	1029	0.0064	2771.6	3867.6	161.90	21.60	32.15	201
142	0.717678	1050	0.0063	2817.1	3931.8	162.36	21.57	32.00	203
144	0.705893	1071	0.0062	2862.3	3995.6	162.80	21.53	31.87	204
146	0.694542	1091	0.0061	2907.4	4059.3	163.24	21.50	31.75	206
148	0.683597	1111	0.0060	2952.4	4122.7	163.67	21.48	31.64	208
150	0.673033	1131	0.0059	2997.2	4185.9	164.10	21.45	31.54	209
152	0.662830	1151	0.0058	3041.9	4248.8	164.52	21.43	31.45	211
154	0.652966	1171	0.0057	3086.5	4311.7	164.93	21.42	31.37	212
156	0.643424	1191	0.0056	3131.0	4374.3	165.33	21.40	31.29	214
158	0.634185	1210	0.0055	3175.4	4436.8	165.73	21.39	31.22	216
160	0.625235	1230	0.0054	3219.7	4499.2	166.12	21.38	31.15	217
165	0.604026	1278	0.0052	3330.1	4654.6	167.08	21.37	31.01	221
170	0.584328	1326	0.0050	3440.3	4809.4	168.00	21.37	30.90	225
175	0.565972	1373	0.0049	3550.2	4963.7	168.90	21.38	30.82	228
180	0.548814	1419	0.0047	3659.9	5117.6	169.76	21.40	30.75	232
185	0.532734	1466	0.0046	3769.5	5271.2	170.60	21.43	30.70	235
190	0.517627	1512	0.0044	3879.1	5424.6	171.42	21.46	30.67	238
195	0.503401	1558	0.0043	3988.7	5577.9	172.22	21.51	30.65	242
200	0.489977	1603	0.0042	4098.4	5731.2	173.00	21.55	30.65	245
205	0.477286	1648	0.0041	4208.3	5884.4	173.75	21.61	30.65	248
210	0.465266	1693	0.0040	4318.3	6037.7	174.49	21.67	30.66	251
215	0.453864	1738	0.0039	4428.4	6191.1	175.21	21.73	30.69	254
220	0.443030	1783	0.0038	4538.8	6344.6	175.92	21.80	30.72	257
225	0.432722	1827	0.0037	4649.5	6498.2	176.61	21.87	30.75	260
230	0.422900	1871	0.0036	4760.4	6652.1	177.29	21.94	30.80	263
235	0.413530	1916	0.0035	4871.7	6806.2	177.95	22.02	30.84	266
240	0.404581	1960	0.0034	4983.2	6960.6	178.60	22.10	30.90	269
245	0.396023	2003	0.0034	5095.1	7115.2	179.24	22.18	30.95	271
250	0.387830	2047	0.0033	5207.3	7270.1	179.86	22.26	31.01	274
255	0.379980	2091	0.0032	5319.9	7425.3	180.48	22.35	31.08	277
260	0.372451	2134	0.0031	5432.9	7580.9	181.08	22.44	31.14	279
265	0.365222	2178	0.0031	5546.5	7736.7	181.67	22.53	31.21	282
270	0.358276	2221	0.0030	5660.1	7893.0	182.26	22.61	31.28	284
275	0.351596	2265	0.0030	5774.2	8049.6	182.83	22.71	31.36	287
280	0.345166	2308	0.0029	5888.8	8206.6	183.40	22.80	31.43	289
285	0.338973	2351	0.0029	6003.8	8363.9	183.96	22.89	31.51	292
290	0.333003	2394	0.0028	6119.3	8521.6	184.50	22.98	31.58	294
295	0.327245	2437	0.0028	6235.1	8679.7	185.04	23.07	31.66	297
300	0.321686	2480	0.0027	6351.4	8838.3	185.58	23.16	31.74	299

* TWO-PHASE BOUNDARY

TABLE 16. THERMODYNAMIC PROPERTIES OF FLUORINE

0.9 MN/m² ISOBAR

TEMPERATURE (IPTS 1968) K	DENSITY MOL/L	ISOTHERM DERIVATIVE J/MOL	ISOCHORE DERIVATIVE MN/m²-K	INTERNAL ENERGY J/MOL	ENTHALPY J/MOL	ENTROPY J/MOL-K	C_v J / MOL - K	C_p J / MOL - K	VELOCITY OF SOUND M/S
* 53.569	44.8962	27113	4.3144	-6045.8	-6025.8	60.85	36.47	54.72	1035
54	44.8258	27056	4.3141	-6022.1	-6002.0	61.29	36.40	54.89	1036
56	44.4990	26519	4.3121	-5911.7	-5891.5	63.30	36.05	55.08	1040
58	44.1751	26109	4.2175	-5801.2	-5780.9	65.25	35.69	55.93	1038
60	43.8598	25888	4.1333	-5691.7	-5671.2	67.11	35.33	55.91	1038
62	43.5363	25527	4.0570	-5581.0	-5560.3	68.93	34.98	56.07	1038
64	43.2037	24676	3.9649	-5469.0	-5448.2	70.71	34.58	56.42	1029
66	42.8704	23575	3.8343	-5356.9	-5335.9	72.44	34.15	56.55	1014
68	42.5509	22846	3.6540	-5246.7	-5225.5	74.09	33.72	55.67	996
70	42.2226	21691	3.5271	-5135.5	-5114.2	75.70	33.32	55.84	978
72	41.8753	20416	3.4178	-5021.9	-5000.4	77.31	32.92	56.41	960
74	41.5365	20048	3.3345	-4909.7	-4888.0	78.85	32.55	56.34	956
76	41.1913	19150	3.2034	-4796.6	-4774.8	80.36	32.14	56.14	938
78	40.8429	18000	3.0834	-4683.4	-4661.4	81.83	31.76	56.46	918
80	40.4895	17130	2.9835	-4569.8	-4547.6	83.27	31.40	56.76	903
82	40.1218	16260	2.8699	-4454.3	-4431.8	84.69	31.02	56.83	885
84	39.7688	15507	2.7577	-4341.1	-4318.5	86.05	30.69	56.74	869
86	39.4013	14576	2.6395	-4226.1	-4203.3	87.40	30.36	56.84	847
88	39.0281	13720	2.5205	-4110.5	-4087.5	88.73	30.05	56.80	826
90	38.6497	12756	2.4443	-3994.6	-3971.3	90.03	29.65	58.07	808
92	38.2631	12122	2.3608	-3877.5	-3854.0	91.31	29.62	58.51	794
94	37.8692	11525	2.2678	-3759.6	-3735.8	92.57	29.37	58.62	778
96	37.4688	10730	2.1706	-3640.9	-3616.9	93.81	29.14	59.17	757
98	37.0589	10107	2.0832	-3521.0	-3496.7	95.03	28.97	59.61	740
100	36.6403	9392	1.9952	-3400.0	-3375.5	96.24	28.77	60.34	720
102	36.2105	8822	1.9105	-3277.6	-3252.7	97.44	28.62	60.80	702
104	35.7692	8256	1.8212	-3153.6	-3128.4	98.62	28.45	61.11	683
106	35.3159	7538	1.7280	-3027.9	-3002.4	99.80	28.31	61.97	659
108	34.8475	6894	1.6436	-2900.3	-2874.4	100.97	28.19	63.04	637
110	34.3624	6294	1.5675	-2770.3	-2744.1	102.14	28.08	64.45	617
* 110.194	34.3143	6238	1.5560	-2757.6	-2731.3	102.25	28.07	64.39	614
* 110.194	1.15345	645	0.0111	2015.6	2795.9	152.37	23.69	39.55	168
112	1.12353	672	0.0107	2065.4	2866.4	153.00	23.43	38.60	171
114	1.09287	701	0.0103	2119.2	2942.7	153.68	23.19	37.72	173
116	1.06446	728	0.0100	2171.9	3017.4	154.33	22.97	36.96	176
118	1.03799	755	0.0097	2223.6	3090.6	154.95	22.78	36.30	178
120	1.01323	781	0.0094	2274.4	3162.7	155.56	22.62	35.73	180
122	0.989969	806	0.0091	2324.5	3233.6	156.15	22.47	35.24	182
124	0.968051	830	0.0088	2373.9	3303.6	156.72	22.34	34.80	184
126	0.947336	854	0.0086	2422.8	3372.8	157.27	22.23	34.41	187
128	0.927706	878	0.0084	2471.2	3441.3	157.81	22.12	34.06	189
130	0.909061	901	0.0082	2519.1	3509.1	158.33	22.03	33.75	191
132	0.891313	924	0.0080	2566.6	3576.3	158.85	21.95	33.47	193
134	0.874388	946	0.0078	2613.7	3643.0	159.35	21.88	33.22	194
136	0.858219	968	0.0077	2660.5	3709.2	159.84	21.82	32.99	196
138	0.842748	990	0.0075	2707.0	3775.0	160.32	21.76	32.78	198
140	0.827922	1012	0.0073	2753.3	3840.3	160.79	21.71	32.60	200
142	0.813696	1033	0.0072	2799.3	3905.4	161.25	21.66	32.43	202
144	0.800028	1054	0.0071	2845.1	3970.1	161.70	21.62	32.27	203
146	0.786882	1075	0.0069	2890.7	4034.5	162.15	21.59	32.13	205
148	0.774223	1096	0.0068	2936.1	4098.6	162.58	21.55	32.00	207
150	0.762022	1116	0.0067	2981.4	4162.4	163.01	21.53	31.88	209
152	0.750250	1137	0.0066	3026.5	4226.1	163.43	21.50	31.77	210
154	0.738883	1157	0.0065	3071.5	4289.5	163.85	21.48	31.67	212
156	0.727897	1177	0.0064	3116.3	4352.8	164.26	21.46	31.58	213
158	0.717271	1197	0.0063	3161.1	4415.8	164.66	21.45	31.49	215
160	0.706986	1217	0.0062	3205.7	4478.7	165.05	21.44	31.41	217
165	0.682649	1266	0.0059	3317.0	4635.4	166.02	21.42	31.25	220
170	0.660087	1314	0.0057	3427.8	4791.2	166.95	21.41	31.11	224
175	0.639097	1362	0.0055	3538.3	4946.5	167.85	21.42	31.01	228
180	0.619505	1410	0.0053	3648.6	5101.3	168.72	21.43	30.92	231
185	0.601166	1457	0.0052	3758.7	5255.8	169.57	21.46	30.86	235
190	0.583955	1503	0.0050	3868.8	5410.0	170.39	21.49	30.82	238
195	0.567764	1550	0.0049	3978.8	5564.0	171.19	21.53	30.79	242
200	0.552500	1596	0.0047	4088.9	5717.9	171.97	21.58	30.77	245
205	0.538081	1641	0.0046	4199.1	5871.7	172.73	21.63	30.77	248
210	0.524435	1687	0.0045	4309.4	6025.6	173.47	21.68	30.77	251
215	0.511498	1732	0.0044	4419.9	6179.5	174.19	21.75	30.79	254
220	0.499213	1777	0.0043	4530.6	6333.5	174.90	21.81	30.81	257
225	0.487530	1822	0.0042	4641.6	6487.6	175.59	21.88	30.84	260
230	0.476405	1866	0.0041	4752.8	6641.9	176.27	21.96	30.88	263
235	0.465797	1911	0.0040	4864.2	6796.4	176.94	22.03	30.92	266
240	0.455668	1955	0.0039	4976.0	6951.2	177.59	22.11	30.97	268
245	0.445986	1999	0.0038	5088.2	7106.2	178.23	22.19	31.03	271
250	0.436722	2043	0.0037	5200.6	7261.4	178.86	22.28	31.08	274
255	0.427847	2087	0.0036	5313.4	7417.0	179.47	22.36	31.14	277
260	0.419338	2131	0.0036	5426.6	7572.9	180.08	22.45	31.21	279
265	0.411170	2175	0.0035	5540.2	7729.1	180.67	22.54	31.27	282
270	0.403324	2218	0.0034	5654.1	7885.6	181.26	22.63	31.34	284
275	0.395781	2262	0.0033	5768.5	8042.5	181.83	22.72	31.41	287
280	0.388522	2305	0.0033	5883.2	8199.7	182.40	22.81	31.49	289
285	0.381531	2349	0.0032	5998.4	8357.3	182.96	22.90	31.56	292
290	0.374795	2392	0.0032	6114.0	8515.3	183.51	22.99	31.63	294
295	0.368297	2435	0.0031	6230.0	8673.7	184.05	23.08	31.71	297
300	0.362027	2478	0.0031	6346.4	8832.4	184.58	23.18	31.79	299

* TWO-PHASE BOUNDARY

TABLE 16. THERMODYNAMIC PROPERTIES OF FLUORINE

1.0 MN/m² ISOBAR

TEMPERATURE (IPTS 1968) K	DENSITY MOL/L	ISOTHERM DERIVATIVE J/MOL	ISOCHORE DERIVATIVE MN/m²-K	INTERNAL ENERGY J/MOL	ENTHALPY J/MOL	ENTROPY J/MOL-K	C_v J / MOL - K	C_p J / MOL - K	VELOCITY OF SOUND M/S
* 53.579	44.8983	27153	4.3149	-6047.9	-6025.7	60.85	36.47	54.70	1035
54	44.8295	27095	4.3145	-6024.8	-6002.4	61.28	36.40	54.86	1037
56	44.5028	26545	4.3128	-5914.4	-5892.0	63.30	36.05	55.86	1040
58	44.1789	26127	4.2190	-5804.0	-5781.3	65.24	35.68	55.93	1038
60	43.8637	25905	4.1341	-5694.5	-5671.7	67.10	35.33	55.90	1039
62	43.5402	25524	4.0567	-5583.8	-5560.8	68.92	34.99	56.08	1038
64	43.2077	24657	3.9644	-5471.9	-5448.7	70.70	34.59	56.44	1029
66	42.8746	23562	3.8353	-5359.8	-5336.4	72.43	34.16	56.57	1013
68	42.5553	22851	3.6958	-5249.6	-5226.1	74.08	33.73	55.69	996
70	42.2272	21702	3.5282	-5138.5	-5114.8	75.69	33.32	55.84	978
72	41.8802	20415	3.4181	-5025.0	-5001.1	77.30	32.93	56.42	959
74	41.5415	20035	3.3343	-4912.8	-4888.7	78.84	32.56	56.36	955
76	41.1966	19144	3.2043	-4799.8	-4775.5	80.35	32.15	56.17	938
78	40.8484	18002	3.0839	-4686.7	-4662.2	81.82	31.77	56.46	918
80	40.4953	17133	2.9840	-4573.1	-4548.4	83.26	31.41	56.77	903
82	40.1279	16254	2.8707	-4457.7	-4432.7	84.68	31.04	56.85	885
84	39.7752	15515	2.7588	-4344.6	-4319.4	86.04	30.70	56.75	869
86	39.4082	14587	2.6409	-4229.6	-4204.3	87.39	30.37	56.85	848
88	39.0354	13731	2.5223	-4114.1	-4088.5	88.72	30.05	56.81	827
90	38.6575	12777	2.4447	-3998.3	-3972.4	90.01	29.87	58.04	808
92	38.2713	12138	2.3612	-3881.4	-3855.2	91.29	29.64	58.49	794
94	37.8779	11537	2.2684	-3763.5	-3737.1	92.55	29.38	58.61	778
96	37.4781	10749	2.1714	-3644.9	-3618.3	93.79	29.16	59.14	757
98	37.0688	10124	2.0839	-3525.2	-3498.2	95.02	28.99	59.58	740
100	36.6509	9412	1.9961	-3404.3	-3377.0	96.23	28.79	60.31	720
102	36.2218	8840	1.9114	-3282.0	-3254.4	97.42	28.64	60.77	703
104	35.7813	8272	1.8226	-3158.1	-3130.2	98.61	28.47	61.09	683
106	35.3291	7561	1.7297	-3032.7	-3004.4	99.78	28.32	61.93	660
108	34.8620	6919	1.6451	-2905.2	-2876.5	100.95	28.20	62.96	638
110	34.3782	6321	1.5694	-2775.5	-2746.4	102.11	28.10	64.36	617
* 111.812	33.9237	5921	1.4928	-2655.8	-2626.3	103.17	28.01	64.57	599
* 111.812	1.27991	635	0.0125	2023.5	2804.8	151.71	23.93	40.65	168
112	1.27623	638	0.0124	2028.9	2812.4	151.78	23.89	40.53	169
114	1.23896	668	0.0119	2085.2	2892.3	152.48	23.59	39.38	171
116	1.20472	698	0.0115	2140.0	2970.0	153.16	23.33	38.41	174
118	1.17305	726	0.0111	2193.5	3046.0	153.81	23.10	37.58	176
120	1.14361	753	0.0107	2246.0	3120.4	154.43	22.90	36.87	179
122	1.11611	780	0.0104	2297.6	3193.5	155.04	22.72	36.25	181
124	1.09031	806	0.0101	2348.3	3265.5	155.62	22.57	35.71	183
126	1.06603	831	0.0098	2398.4	3336.4	156.19	22.43	35.23	185
128	1.04310	855	0.0095	2447.8	3406.5	156.74	22.31	34.81	187
130	1.02139	879	0.0093	2496.6	3475.7	157.28	22.20	34.44	189
132	1.00079	903	0.0091	2545.0	3544.2	157.80	22.11	34.10	191
134	0.981200	926	0.0089	2593.0	3612.1	158.31	22.02	33.80	193
136	0.962525	949	0.0087	2640.5	3679.5	158.81	21.94	33.53	195
138	0.944695	972	0.0085	2687.8	3746.3	159.30	21.88	33.29	197
140	0.927643	994	0.0083	2734.6	3812.6	159.78	21.82	33.07	199
142	0.911310	1016	0.0081	2781.3	3878.6	160.24	21.76	32.87	201
144	0.895645	1038	0.0080	2827.6	3944.1	160.70	21.71	32.68	203
146	0.880599	1059	0.0078	2873.7	4009.3	161.15	21.67	32.52	204
148	0.866133	1080	0.0077	2919.6	4074.2	161.59	21.63	32.36	206
150	0.852207	1101	0.0075	2965.4	4138.8	162.03	21.60	32.22	208
152	0.838788	1122	0.0074	3010.9	4203.1	162.45	21.57	32.09	210
154	0.825845	1143	0.0073	3056.3	4267.2	162.87	21.55	31.98	211
156	0.813350	1163	0.0071	3101.5	4331.0	163.28	21.53	31.87	213
158	0.801276	1184	0.0070	3146.6	4394.7	163.69	21.51	31.77	215
160	0.789600	1204	0.0069	3191.6	4458.1	164.09	21.49	31.68	216
165	0.762014	1254	0.0066	3303.7	4616.0	165.06	21.46	31.48	220
170	0.736488	1303	0.0064	3415.2	4773.0	166.00	21.45	31.33	224
175	0.712778	1352	0.0062	3526.3	4929.3	166.90	21.45	31.20	227
180	0.690680	1400	0.0060	3637.2	5085.0	167.78	21.46	31.10	231
185	0.670022	1448	0.0058	3747.8	5240.3	168.63	21.49	31.02	235
190	0.650656	1495	0.0056	3858.4	5395.3	169.46	21.52	30.96	238
195	0.632457	1542	0.0054	3968.9	5550.0	170.26	21.55	30.92	241
200	0.615315	1588	0.0053	4079.4	5704.5	171.04	21.60	30.90	245
205	0.599134	1634	0.0051	4189.9	5859.0	171.81	21.65	30.88	248
210	0.583832	1680	0.0050	4300.6	6013.4	172.55	21.70	30.88	251
215	0.569334	1726	0.0049	4411.4	6167.8	173.28	21.76	30.89	254
220	0.555576	1771	0.0048	4522.4	6322.3	173.99	21.83	30.91	257
225	0.542500	1817	0.0046	4633.6	6476.9	174.68	21.90	30.93	260
230	0.530054	1861	0.0045	4745.1	6631.7	175.36	21.97	30.97	263
235	0.518191	1906	0.0044	4856.8	6786.6	176.03	22.05	31.01	266
240	0.506870	1951	0.0043	4968.9	6941.8	176.68	22.13	31.05	268
245	0.496052	1995	0.0042	5081.2	7097.1	177.32	22.21	31.10	271
250	0.485705	2040	0.0041	5193.9	7252.7	177.95	22.29	31.15	274
255	0.475796	2084	0.0040	5306.9	7408.7	178.57	22.38	31.21	277
260	0.466297	2128	0.0040	5420.3	7564.9	179.18	22.46	31.27	279
265	0.457184	2172	0.0039	5534.1	7721.4	179.77	22.55	31.33	282
270	0.448431	2216	0.0038	5648.2	7878.2	180.36	22.64	31.40	284
275	0.440017	2259	0.0037	5762.7	8035.4	180.94	22.73	31.47	287
280	0.431923	2303	0.0037	5877.7	8192.9	181.50	22.82	31.54	289
285	0.424131	2346	0.0036	5993.0	8350.8	182.06	22.91	31.61	292
290	0.416622	2390	0.0035	6108.8	8509.0	182.61	23.01	31.69	294
295	0.409382	2433	0.0035	6224.9	8667.6	183.16	23.10	31.76	297
300	0.402396	2476	0.0034	6341.5	8826.6	183.69	23.19	31.84	299

* TWO-PHASE BOUNDARY

TABLE 16. THERMODYNAMIC PROPERTIES OF FLUORINE

1.2 MN/m² ISOBAR

TEMPERATURE (IPTS 1968) K	DENSITY MOL/L	ISOTHERM DERIVATIVE J/MOL	ISOCHORE DERIVATIVE MN/m²-K	INTERNAL ENERGY J/MOL	ENTHALPY J/MOL	ENTROPY J/MOL-K	C_v J / MOL - K	C_p J / MOL - K	VELOCITY OF SOUND M/S
* 53.598	44.9024	27234	4.3158	-6052.1	-6025.4	60.86	36.46	54.65	1036
54	44.8372	27168	4.3154	-6030.1	-6003.3	61.27	36.40	54.81	1038
56	44.5103	26598	4.3136	-5919.8	-5892.9	63.28	36.05	55.82	1041
58	44.1866	26162	4.2219	-5809.5	-5782.3	65.22	35.68	55.92	1039
60	43.8714	25938	4.1357	-5700.1	-5672.7	67.08	35.33	55.89	1039
62	43.5481	25520	4.0561	-5589.4	-5561.9	68.90	35.00	56.08	1037
64	43.2159	24619	3.9636	-5477.6	-5449.8	70.69	34.61	56.47	1028
66	42.8831	23538	3.8372	-5365.6	-5337.6	72.41	34.17	56.63	1013
68	42.5641	22860	3.6593	-5255.5	-5227.3	74.06	33.74	55.72	997
70	42.2364	21722	3.5302	-5144.5	-5116.1	75.68	33.33	55.84	979
72	41.8899	20414	3.4188	-5031.1	-5002.5	77.28	32.94	56.43	959
74	41.5515	20010	3.3340	-4919.0	-4890.1	78.82	32.58	56.39	955
76	41.2070	19132	3.2060	-4806.1	-4777.0	80.33	32.17	56.21	938
78	40.8595	18006	3.0849	-4693.2	-4663.8	81.80	31.79	56.48	918
80	40.5070	17139	2.9849	-4579.7	-4550.1	83.23	31.43	56.78	903
82	40.1403	16242	2.8722	-4464.4	-4434.5	84.66	31.06	56.90	885
84	39.7881	15530	2.7610	-4351.5	-4321.3	86.02	30.72	56.77	869
86	39.4219	14609	2.6437	-4236.7	-4206.3	87.37	30.39	56.86	848
88	39.0500	13754	2.5259	-4121.4	-4090.6	88.69	30.06	56.83	827
90	38.6732	12820	2.4455	-4005.8	-3974.7	89.99	29.90	57.97	809
92	38.2878	12171	2.3620	-3889.0	-3857.7	91.27	29.68	58.44	794
94	37.8952	11562	2.2696	-3771.3	-3739.7	92.53	29.42	58.58	778
96	37.4967	10786	2.1730	-3653.0	-3621.0	93.76	29.19	59.09	758
98	37.0885	10160	2.0853	-3533.5	-3501.1	94.99	29.03	59.52	740
100	36.6721	9454	1.9980	-3412.9	-3380.2	96.19	28.83	60.23	721
102	36.2444	8876	1.9131	-3290.9	-3257.8	97.39	28.69	60.70	703
104	35.8054	8305	1.8253	-3167.2	-3133.7	98.57	28.51	61.06	684
106	35.3555	7607	1.7332	-3042.2	-3008.2	99.74	28.36	61.84	661
108	34.8908	6970	1.6481	-2915.1	-2880.7	100.91	28.23	62.80	639
110	34.4097	6375	1.5733	-2785.9	-2751.0	102.07	28.13	64.20	619
112	33.9095	5923	1.4898	-2654.0	-2618.6	103.23	28.03	64.53	599
114	33.3894	5264	1.4116	-2519.4	-2483.4	104.40	27.94	66.64	575
* 114.725	33.1947	5094	1.3816	-2469.7	-2433.6	104.82	27.91	66.92	567
* 114.725	1.53615	612	0.0153	2032.9	2814.1	150.53	24.38	42.96	168
116	1.50529	633	0.0148	2071.0	2868.2	151.00	24.14	41.97	170
118	1.46051	666	0.0142	2129.2	2950.8	151.71	23.81	40.65	173
120	1.41952	696	0.0137	2185.6	3031.0	152.38	23.52	39.55	176
122	1.38173	726	0.0132	2240.6	3109.1	153.03	23.28	38.61	178
124	1.34668	755	0.0128	2294.4	3185.5	153.65	23.06	37.81	180
126	1.31400	782	0.0124	2347.2	3260.4	154.25	22.87	37.11	183
128	1.28341	809	0.0120	2399.0	3334.0	154.83	22.71	36.51	185
130	1.25467	835	0.0116	2450.1	3406.5	155.39	22.56	35.98	187
132	1.22757	861	0.0113	2500.4	3478.0	155.93	22.43	35.51	189
134	1.20194	886	0.0110	2550.2	3548.6	156.46	22.32	35.09	191
136	1.17765	910	0.0108	2599.4	3618.4	156.98	22.21	34.72	194
138	1.15456	934	0.0105	2648.1	3687.5	157.49	22.12	34.38	195
140	1.13258	958	0.0103	2696.4	3755.9	157.98	22.04	34.08	197
142	1.11161	981	0.0101	2744.3	3823.8	158.46	21.97	33.81	199
144	1.09156	1004	0.0098	2791.8	3891.2	158.93	21.91	33.57	201
146	1.07238	1027	0.0096	2839.1	3958.1	159.39	21.85	33.34	203
148	1.05399	1049	0.0094	2886.0	4024.6	159.85	21.80	33.14	205
150	1.03633	1071	0.0093	2932.7	4090.7	160.29	21.75	32.95	207
152	1.01936	1093	0.0091	2979.2	4156.4	160.72	21.72	32.78	208
154	1.00304	1115	0.0089	3025.4	4221.8	161.15	21.68	32.63	210
156	0.987309	1136	0.0088	3071.5	4286.9	161.57	21.65	32.49	212
158	0.972144	1157	0.0086	3117.4	4351.7	161.98	21.62	32.35	213
160	0.957508	1178	0.0085	3163.1	4416.3	162.39	21.60	32.23	215
165	0.923034	1230	0.0081	3276.8	4576.8	163.38	21.56	31.98	219
170	0.891261	1281	0.0078	3389.8	4736.2	164.33	21.54	31.77	223
175	0.861848	1331	0.0075	3502.2	4894.6	165.25	21.53	31.60	227
180	0.834516	1361	0.0073	3614.2	5052.2	166.14	21.53	31.46	230
185	0.809031	1430	0.0070	3725.9	5209.2	167.00	21.54	31.35	234
190	0.785195	1478	0.0068	3837.4	5365.7	167.83	21.57	31.27	237
195	0.762841	1526	0.0066	3948.8	5521.9	168.64	21.60	31.20	241
200	0.741824	1574	0.0064	4060.1	5677.8	169.43	21.64	31.15	244
205	0.722018	1621	0.0062	4171.4	5833.5	170.20	21.69	31.12	247
210	0.703315	1667	0.0061	4282.8	5989.0	170.95	21.74	31.11	251
215	0.685619	1714	0.0059	4394.3	6144.5	171.68	21.80	31.10	254
220	0.668847	1760	0.0058	4505.9	6300.0	172.40	21.86	31.10	257
225	0.652923	1806	0.0056	4617.7	6455.6	173.10	21.93	31.12	260
230	0.637782	1852	0.0055	4729.7	6611.2	173.78	22.00	31.14	263
235	0.623363	1897	0.0053	4842.0	6767.0	174.45	22.08	31.17	265
240	0.609615	1942	0.0052	4954.5	6922.9	175.11	22.16	31.20	268
245	0.596489	1987	0.0051	5067.3	7079.1	175.75	22.24	31.25	271
250	0.583942	2032	0.0050	5180.4	7235.4	176.38	22.32	31.29	274
255	0.571936	2077	0.0049	5293.8	7392.0	177.00	22.40	31.34	277
260	0.560433	2121	0.0048	5407.6	7548.8	177.61	22.49	31.40	279
265	0.549403	2166	0.0047	5521.8	7706.0	178.21	22.58	31.46	282
270	0.538815	2210	0.0046	5636.3	7863.4	178.80	22.67	31.52	284
275	0.528644	2254	0.0045	5751.2	8021.2	179.38	22.76	31.58	287
280	0.518862	2298	0.0044	5866.5	8179.2	179.95	22.85	31.65	289
285	0.509449	2342	0.0043	5982.2	8337.6	180.51	22.94	31.72	292
290	0.500383	2386	0.0042	6098.2	8496.4	181.06	23.03	31.79	294
295	0.491644	2429	0.0042	6214.7	8655.5	181.61	23.12	31.86	297
300	0.483215	2473	0.0041	6331.6	8815.0	182.14	23.22	31.93	299

* TWO-PHASE BOUNDARY

TABLE 16. THERMODYNAMIC PROPERTIES OF FLUORINE

1.4 MN/m² ISOBAR

TEMPERATURE (IPTS 1968) K	DENSITY MOL/L	ISOTHERM DERIVATIVE J/MOL	ISOCHORE DERIVATIVE MN/m²-K	INTERNAL ENERGY J/MOL	ENTHALPY J/MOL	ENTROPY J/MOL-K	Cv J / MOL - K	Cp J / MOL - K	VELOCITY OF SOUND M/S
* 53.618	44.9065	27313	4.3167	-6056.3	-6025.2	60.86	36.46	54.60	1038
54	44.8445	27246	4.3163	-6035.4	-6004.2	61.25	36.39	54.75	1039
56	44.5179	26650	4.3144	-5925.2	-5893.8	63.26	36.04	55.78	1042
58	44.1942	26197	4.2249	-5814.9	-5783.3	65.21	35.68	55.91	1039
60	43.8791	25972	4.1374	-5705.6	-5673.7	67.07	35.33	55.87	1040
62	43.5559	25516	4.0556	-5595.0	-5562.9	68.89	35.02	56.08	1037
64	43.2240	24583	3.9628	-5483.3	-5450.9	70.67	34.62	56.50	1028
66	42.8916	23515	3.8390	-5371.4	-5338.8	72.40	34.19	56.68	1013
68	42.5728	22870	3.6628	-5261.4	-5228.5	74.04	33.74	55.75	997
70	42.2456	21744	3.5322	-5150.5	-5117.3	75.66	33.34	55.84	979
72	41.8997	20415	3.4196	-5037.3	-5003.8	77.26	32.95	56.44	959
74	41.5615	19987	3.3338	-4925.2	-4891.5	78.80	32.60	56.43	954
76	41.2175	19121	3.2077	-4812.5	-4778.5	80.31	32.19	56.26	938
78	40.8706	18011	3.0859	-4699.6	-4665.4	81.78	31.80	56.49	918
80	40.5187	17145	2.9857	-4586.3	-4551.8	83.21	31.46	56.79	903
82	40.1526	16232	2.8737	-4471.2	-4436.3	84.64	31.08	56.95	885
84	39.8010	15546	2.7631	-4358.4	-4323.2	86.00	30.74	56.78	869
86	39.4356	14632	2.6464	-4243.8	-4208.2	87.35	30.40	56.87	849
88	39.0645	13778	2.5294	-4128.6	-4092.7	88.67	30.07	56.85	828
90	38.6887	12862	2.4464	-4013.2	-3977.0	89.96	29.93	57.91	809
92	38.3042	12203	2.3628	-3896.6	-3860.1	91.24	29.71	58.40	795
94	37.9125	11588	2.2708	-3779.1	-3742.2	92.50	29.46	58.56	779
96	37.5152	10823	2.1750	-3661.1	-3623.7	93.74	29.23	59.04	759
98	37.1082	10196	2.0867	-3541.8	-3504.0	94.96	29.07	59.47	741
100	36.6932	9495	1.9998	-3421.4	-3383.3	96.16	28.87	60.16	722
102	36.2668	8913	1.9150	-3299.7	-3261.1	97.36	28.73	60.63	704
104	35.8295	8338	1.8280	-3176.3	-3137.2	98.54	28.55	61.02	685
106	35.3817	7654	1.7365	-3051.6	-3012.0	99.71	28.39	61.75	662
108	34.9194	7021	1.6511	-2925.0	-2884.9	100.87	28.26	62.65	640
110	34.4410	6428	1.5771	-2796.2	-2755.5	102.03	28.17	64.05	620
112	33.9432	5962	1.4937	-2664.7	-2623.4	103.19	28.05	64.43	600
114	33.4272	5327	1.4164	-2530.7	-2488.8	104.35	27.96	66.38	577
116	32.8858	4815	1.3351	-2393.3	-2350.7	105.51	27.89	67.59	554
* 117.306	32.5174	4439	1.2858	-2301.6	-2258.5	106.27	27.85	69.16	539
* 117.306	1.79777	587	0.0183	2035.8	2814.5	149.50	24.82	45.41	168
118	1.77662	600	0.0179	2057.8	2845.8	149.76	24.67	44.72	169
120	1.71988	635	0.0171	2119.4	2933.4	150.50	24.26	42.98	172
122	1.66848	669	0.0164	2178.8	3017.9	151.20	23.92	41.56	175
124	1.62150	701	0.0158	2236.4	3099.8	151.87	23.63	40.38	178
126	1.57824	732	0.0152	2292.5	3179.5	152.50	23.37	39.37	180
128	1.53817	761	0.0147	2347.2	3257.4	153.12	23.15	38.51	183
130	1.50086	790	0.0142	2400.9	3333.7	153.71	22.96	37.77	185
132	1.46595	817	0.0138	2453.5	3408.5	154.28	22.79	37.12	187
134	1.43318	844	0.0134	2505.4	3482.2	154.83	22.64	36.55	189
136	1.40230	870	0.0131	2556.4	3554.8	155.37	22.51	36.05	192
138	1.37311	896	0.0127	2606.9	3626.5	155.89	22.39	35.61	194
140	1.34546	921	0.0124	2656.7	3697.3	156.40	22.28	35.21	196
142	1.31919	946	0.0121	2706.1	3767.3	156.90	22.19	34.85	198
144	1.29419	970	0.0118	2754.9	3836.7	157.39	22.11	34.53	200
146	1.27034	994	0.0116	2803.4	3905.5	157.86	22.04	34.24	202
148	1.24756	1017	0.0113	2851.5	3973.7	158.32	21.97	33.98	203
150	1.22576	1041	0.0111	2899.2	4041.4	158.78	21.91	33.74	205
152	1.20486	1063	0.0109	2946.7	4108.6	159.22	21.86	33.52	207
154	1.18481	1086	0.0107	2993.9	4175.5	159.66	21.82	33.32	209
156	1.16554	1108	0.0105	3040.8	4241.9	160.09	21.78	33.14	211
158	1.14700	1130	0.0103	3087.5	4308.1	160.51	21.74	32.97	212
160	1.12914	1152	0.0101	3134.0	4373.9	160.92	21.71	32.82	214
165	1.08722	1206	0.0097	3249.4	4537.1	161.93	21.66	32.49	218
170	1.04875	1259	0.0093	3364.0	4698.9	162.89	21.62	32.23	222
175	1.01325	1311	0.0089	3477.8	4859.4	163.83	21.60	32.01	226
180	0.980376	1362	0.0086	3591.0	5019.0	164.72	21.59	31.83	230
185	0.949802	1412	0.0083	3703.8	5177.8	165.60	21.60	31.69	233
190	0.921274	1461	0.0081	3816.3	5336.0	166.44	21.62	31.57	237
195	0.894575	1510	0.0078	3928.6	5493.6	167.26	21.65	31.49	240
200	0.869520	1559	0.0076	4040.8	5650.8	168.05	21.69	31.42	244
205	0.845949	1607	0.0074	4152.9	5807.8	168.83	21.73	31.37	247
210	0.823724	1655	0.0071	4264.9	5964.5	169.58	21.78	31.33	250
215	0.802724	1702	0.0070	4377.1	6121.1	170.32	21.84	31.31	253
220	0.782845	1749	0.0068	4489.3	6277.7	171.04	21.90	31.30	256
225	0.763992	1795	0.0066	4601.7	6434.2	171.74	21.96	31.30	260
230	0.746084	1842	0.0064	4714.3	6590.7	172.43	22.03	31.32	262
235	0.729047	1888	0.0063	4827.0	6747.3	173.11	22.11	31.33	265
240	0.712816	1934	0.0061	4940.0	6904.1	173.77	22.19	31.36	268
245	0.697332	1979	0.0060	5053.3	7061.0	174.41	22.26	31.39	271
250	0.682542	2025	0.0059	5166.9	7218.0	175.05	22.35	31.43	274
255	0.668398	2070	0.0057	5280.8	7375.3	175.67	22.43	31.48	276
260	0.654856	2115	0.0056	5395.0	7532.8	176.28	22.52	31.53	279
265	0.641878	2160	0.0055	5509.5	7690.6	176.88	22.61	31.58	282
270	0.629427	2204	0.0054	5624.4	7848.6	177.47	22.69	31.64	284
275	0.617471	2249	0.0053	5739.6	8007.0	178.06	22.78	31.69	287
280	0.605980	2293	0.0052	5855.3	8165.6	178.63	22.87	31.76	289
285	0.594926	2338	0.0051	5971.3	8324.5	179.19	22.97	31.82	292
290	0.584283	2382	0.0050	6087.7	8483.8	179.74	23.06	31.89	294
295	0.574029	2426	0.0049	6204.5	8643.4	180.29	23.15	31.95	297
300	0.564142	2470	0.0048	6321.7	8803.3	180.83	23.24	32.02	299

* TWO-PHASE BOUNDARY

TABLE 16. THERMODYNAMIC PROPERTIES OF FLUORINE

ISOTHERM DERIVATIVE J/MOL	ISOCHORE DERIVATIVE MN/m²-K	INTERNAL ENERGY J/MOL	ENTHALPY J/MOL	ENTROPY J/MOL-K	C_v J / MOL - K	C_p	VELOCITY OF SOUND M/S
27392	4.3175	-6060.6	-6024.9	60.87	36.45	54.55	1039
27324	4.3172	-6040.7	-6005.0	61.24	36.39	54.70	1040
26703	4.3152	-5930.6	-5894.7	63.25	36.04	55.74	1043
26233	4.2278	-5820.4	-5784.2	65.19	35.67	55.90	1040
26005	4.1393	-5711.2	-5674.7	67.05	33.34	55.86	1040
25513	4.0552	-5600.7	-5563.9	68.87	35.03	56.09	1037
24549	3.9620	-5489.0	-5452.0	70.65	34.64	56.54	1027
23493	3.8408	-5377.2	-5339.9	72.38	34.21	56.73	1013
22881	3.6661	-5267.3	-5229.7	74.03	33.75	55.78	998
21765	3.5343	-5156.5	-5118.6	75.64	33.35	55.85	979
20416	3.4205	-5043.4	-5005.2	77.24	32.96	56.45	959
19964	3.3337	-4931.5	-4893.0	78.78	32.62	56.46	954
19111	3.2094	-4818.8	-4780.0	80.29	32.20	56.30	938
18017	3.0869	-4706.1	-4667.0	81.75	31.82	56.50	918
17153	2.9866	-4592.9	-4553.5	83.19	31.48	56.80	903
16223	2.8752	-4477.9	-4438.1	84.61	31.10	57.00	885
15562	2.7652	-4365.2	-4325.0	85.97	30.76	56.80	870
14655	2.6491	-4250.8	-4210.2	87.32	30.42	56.88	849
13802	2.5329	-4135.8	-4094.8	88.65	30.08	56.87	829
12904	2.4475	-4020.6	-3979.3	89.94	29.96	57.85	810
12237	2.3638	-3904.3	-3862.5	91.21	29.75	58.36	795
11614	2.2720	-3786.9	-3744.7	92.47	29.50	58.54	779
10861	2.1771	-3669.1	-3626.5	93.71	29.26	59.00	759
10232	2.0883	-3550.0	-3506.9	94.93	29.12	59.42	741
9536	2.0017	-3430.0	-3386.4	96.13	28.91	60.08	722
8950	1.9169	-3308.5	-3264.4	97.32	28.77	60.57	704
8371	1.8307	-3185.4	-3140.8	98.50	28.59	60.98	685
7700	1.7398	-3061.0	-3015.8	99.67	28.42	61.66	663
7072	1.6541	-2934.8	-2889.0	100.83	28.29	62.50	641
6482	1.5808	-2806.4	-2760.0	101.99	28.20	63.89	622
6003	1.4975	-2675.3	-2628.2	103.15	28.08	64.32	602
5390	1.4211	-2542.0	-2494.2	104.30	27.99	66.13	579
4875	1.3401	-2405.2	-2356.6	105.46	27.91	67.32	556
4303	1.2628	-2264.8	-2215.4	106.63	27.84	69.60	532
3909	1.2012	-2146.8	-2096.6	107.59	27.81	71.26	513
562	0.0214	2033.2	2807.8	148.57	25.25	48.05	168
569	0.0212	2045.6	2825.3	148.71	25.15	47.59	168
608	0.0201	2110.8	2918.2	149.48	24.68	45.38	171
644	0.0192	2173.3	3007.1	150.20	24.29	43.60	174
678	0.0184	2233.5	3092.8	150.89	23.95	42.14	177
711	0.0177	2291.8	3175.8	151.54	23.66	40.92	180
742	0.0171	2348.6	3256.6	152.17	23.40	39.89	182
772	0.0165	2404.0	3335.5	152.77	23.18	39.00	185
801	0.0160	2458.2	3412.7	153.35	22.99	38.24	187
829	0.0155	2511.5	3488.5	153.91	22.82	37.57	190
857	0.0151	2563.9	3563.0	154.46	22.68	36.99	192
884	0.0147	2615.5	3636.5	154.99	22.54	36.47	194
910	0.0143	2666.4	3708.9	155.50	22.43	36.01	196
935	0.0140	2716.8	3780.5	156.00	22.33	35.59	198
961	0.0136	2766.6	3851.3	156.49	22.23	35.22	200
985	0.0133	2815.9	3921.4	156.97	22.15	34.89	202
1010	0.0130	2864.8	3990.9	157.43	22.08	34.59	204
1034	0.0128	2913.4	4059.8	157.89	22.02	34.31	206
1057	0.0125	2961.5	4128.2	158.34	21.96	34.07	208
1080	0.0123	3009.4	4196.1	158.77	21.91	33.84	210
1103	0.0120	3057.0	4263.6	159.20	21.87	33.63	211
1126	0.0118	3104.3	4330.6	159.62	21.83	33.44	213
1182	0.0113	3221.6	4496.8	160.65	21.75	33.04	217
1236	0.0108	3337.8	4661.1	161.63	21.70	32.70	221
1290	0.0104	3453.0	4823.9	162.57	21.67	32.44	225
1342	0.0100	3567.5	4985.6	163.48	21.66	32.22	229
1394	0.0097	3681.5	5146.2	164.36	21.66	32.04	233
1445	0.0093	3795.0	5306.0	165.22	21.67	31.89	237
1495	0.0090	3908.2	5465.1	166.04	21.70	31.78	240
1544	0.0088	4021.3	5623.8	166.85	21.73	31.69	243
1593	0.0085	4134.1	5782.0	167.63	21.77	31.61	247
1642	0.0082	4247.0	5940.0	168.39	21.82	31.56	250
1690	0.0080	4359.8	6097.7	169.13	21.87	31.53	253
1738	0.0078	4472.7	6255.2	169.86	21.93	31.50	256
1785	0.0076	4585.6	6412.7	170.56	22.00	31.49	259
1832	0.0074	4698.8	6570.2	171.26	22.07	31.49	262
1879	0.0072	4812.1	6727.7	171.93	22.14	31.50	265
1925	0.0071	4925.6	6885.2	172.60	22.21	31.52	268
1971	0.0069	5039.3	7042.9	173.25	22.29	31.54	271
2017	0.0067	5153.3	7200.7	173.88	22.37	31.58	274
2063	0.0066	5267.6	7358.6	174.51	22.46	31.61	276
2109	0.0064	5382.2	7516.8	175.12	22.54	31.66	279
2154	0.0063	5497.2	7675.2	175.73	22.63	31.70	282
2199	0.0062	5612.4	7833.9	176.32	22.72	31.75	284
2244	0.0061	5728.1	7992.8	176.90	22.81	31.81	287
2289	0.0059	5844.1	8151.9	177.48	22.90	31.87	290
2333	0.0058	5960.4	8311.4	178.04	22.99	31.93	292
2378	0.0057	6077.1	8471.2	178.60	23.08	31.99	294
2422	0.0056	6194.3	8631.3	179.14	23.18	32.05	297
2467	0.0055	6311.8	8791.7	179.68	23.27	32.12	299

TABLE 16. THERMODYNAMIC PROPERTIES OF FLUORINE

1.8 MN/m² ISOBAR

TEMPERATURE (IPTS 1968) K	DENSITY MOL/L	ISOTHERM DERIVATIVE J/MOL	ISOCHORE DERIVATIVE MN/m²-K	INTERNAL ENERGY J/MOL	ENTHALPY J/MOL	ENTROPY J/MOL-K	C_v J / MOL - K	C_p J / MOL - K	VELOCITY OF SOUND M/S
* 53.657	44.9148	27471	4.3184	-6064.8	-6024.7	60.87	36.44	54.50	1040
54	44.8592	27402	4.3180	-6046.0	-6005.9	61.22	36.39	54.64	1041
56	44.5328	26755	4.3160	-5936.0	-5895.6	63.23	36.04	55.70	1043
58	44.2094	26268	4.2307	-5825.9	-5785.1	65.17	35.67	55.89	1041
60	43.8945	26039	4.1412	-5716.7	-5675.7	67.03	35.34	55.85	1041
62	43.5716	25510	4.0549	-5606.3	-5565.0	68.85	35.04	56.09	1037
64	43.2403	24516	3.9612	-5494.8	-5453.1	70.63	34.66	56.57	1026
66	42.9086	23473	3.8425	-5383.0	-5341.1	72.36	34.22	56.77	1012
68	42.5903	22892	3.6694	-5273.2	-5230.9	74.01	33.76	55.81	998
70	42.2640	21787	3.5363	-5162.5	-5119.9	75.62	33.35	55.85	980
72	41.9193	20418	3.4214	-5049.5	-5006.6	77.22	32.97	56.46	959
74	41.5815	19944	3.3336	-4937.7	-4894.4	78.76	32.64	56.49	953
76	41.2384	19102	3.2110	-4825.1	-4781.5	80.27	32.22	56.34	938
78	40.8928	18024	3.0880	-4712.6	-4668.6	81.73	31.84	56.51	918
80	40.5420	17161	2.9874	-4599.5	-4555.1	83.17	31.50	56.81	903
82	40.1772	16215	2.8767	-4484.7	-4439.9	84.59	31.12	57.04	884
84	39.8267	15578	2.7673	-4372.1	-4326.9	85.95	30.78	56.81	870
86	39.4629	14678	2.6517	-4257.8	-4212.2	87.30	30.43	56.89	850
88	39.0935	13827	2.5363	-4143.0	-4096.9	88.62	30.10	56.89	829
90	38.7197	12946	2.4488	-4028.1	-3981.6	89.91	29.98	57.79	810
92	38.3369	12270	2.3648	-3911.9	-3864.9	91.19	29.78	58.31	795
94	37.9469	11641	2.2732	-3794.7	-3747.3	92.45	29.53	58.51	779
96	37.5520	10899	2.1792	-3677.1	-3629.2	93.68	29.29	58.95	760
98	37.1473	10268	2.0899	-3558.3	-3509.8	94.90	29.16	59.37	742
100	36.7352	9577	2.0035	-3438.5	-3389.5	96.10	28.95	60.00	723
102	36.3115	8988	1.9188	-3317.2	-3267.7	97.29	28.81	60.50	705
104	35.8773	8405	1.8333	-3194.4	-3144.3	98.47	28.63	60.94	686
106	35.4336	7746	1.7431	-3070.4	-3019.6	99.64	28.45	61.56	664
108	34.9759	7123	1.6572	-2944.6	-2893.1	100.80	28.31	62.36	643
110	34.5027	6535	1.5845	-2816.6	-2764.5	101.95	28.23	63.73	623
112	34.0098	6045	1.5016	-2685.9	-2633.0	103.10	28.11	64.22	603
114	33.5014	5451	1.4257	-2553.2	-2499.5	104.25	28.02	65.89	581
116	32.9679	4934	1.3451	-2417.0	-2362.4	105.41	27.92	67.06	558
118	32.4090	4373	1.2686	-2277.4	-2221.8	106.57	27.86	69.20	535
120	31.8169	3876	1.1909	-2133.3	-2076.8	107.74	27.81	71.19	511
* 121.760	31.2639	3478	1.1253	-2002.4	-1944.8	108.79	27.79	73.14	491
* 121.760	2.34133	535	0.0247	2026.2	2795.0	147.71	25.66	50.94	167
122	2.33037	540	0.0245	2034.7	2807.2	147.81	25.59	50.56	168
124	2.24541	582	0.0232	2103.8	2905.4	148.61	25.06	47.81	171
126	2.17012	621	0.0221	2169.3	2998.8	149.35	24.61	45.64	174
128	2.10252	657	0.0211	2232.1	3088.2	150.06	24.23	43.89	177
130	2.04116	692	0.0203	2292.7	3174.5	150.73	23.90	42.44	180
132	1.98499	725	0.0195	2351.3	3258.2	151.37	23.63	41.23	182
134	1.93321	756	0.0188	2408.5	3339.6	151.98	23.38	40.20	185
136	1.88519	787	0.0182	2464.2	3419.1	152.57	23.17	39.32	187
138	1.84044	816	0.0177	2518.9	3496.9	153.13	22.99	38.55	190
140	1.79854	845	0.0171	2572.5	3573.3	153.68	22.82	37.88	192
142	1.75918	873	0.0167	2625.3	3648.5	154.22	22.68	37.29	194
144	1.72207	900	0.0162	2677.3	3722.5	154.73	22.56	36.77	196
146	1.68698	927	0.0158	2728.6	3795.6	155.24	22.44	36.30	199
148	1.65371	953	0.0154	2779.3	3867.7	155.73	22.34	35.88	201
150	1.62211	978	0.0151	2829.5	3939.1	156.21	22.26	35.51	203
152	1.59201	1003	0.0147	2879.2	4009.8	156.68	22.18	35.17	205
154	1.56329	1028	0.0144	2928.4	4079.8	157.13	22.11	34.86	207
156	1.53585	1052	0.0141	2977.3	4149.3	157.58	22.05	34.59	208
158	1.50958	1076	0.0138	3025.8	4218.2	158.02	22.00	34.33	210
160	1.48439	1100	0.0136	3074.0	4286.6	158.45	21.95	34.10	212
165	1.42568	1158	0.0129	3193.3	4455.9	159.49	21.86	33.61	216
170	1.37228	1214	0.0124	3311.2	4622.9	160.49	21.79	33.21	221
175	1.32339	1269	0.0119	3427.9	4788.0	161.45	21.75	32.88	225
180	1.27840	1323	0.0114	3543.8	4951.8	162.37	21.73	32.61	229
185	1.23680	1376	0.0110	3658.9	5114.3	163.26	21.72	32.40	232
190	1.19818	1428	0.0106	3773.5	5275.8	164.12	21.73	32.22	236
195	1.16219	1479	0.0103	3887.7	5436.5	164.96	21.75	32.07	240
200	1.12854	1530	0.0100	4001.6	5596.6	165.77	21.77	31.96	243
205	1.09700	1580	0.0097	4115.3	5756.1	166.56	21.81	31.87	246
210	1.06735	1629	0.0094	4228.9	5915.3	167.32	21.86	31.80	250
215	1.03942	1678	0.0091	4342.4	6074.1	168.07	21.91	31.74	253
220	1.01304	1727	0.0089	4455.9	6232.8	168.80	21.97	31.71	256
225	0.988076	1775	0.0086	4569.5	6391.2	159.51	22.03	31.68	259
230	0.964416	1822	0.0084	4683.2	6549.6	170.21	22.10	31.67	262
235	0.941949	1870	0.0082	4797.0	6708.0	170.89	22.17	31.67	265
240	0.920581	1917	0.0080	4911.0	6866.3	171.56	22.24	31.68	268
245	0.900229	1964	0.0078	5025.3	7024.8	172.21	22.32	31.70	271
250	0.880818	2010	0.0076	5139.7	7183.3	172.85	22.40	31.72	274
255	0.862279	2056	0.0075	5254.5	7342.0	173.48	22.48	31.75	276
260	0.844553	2102	0.0073	5369.5	7500.8	174.10	22.57	31.79	279
265	0.827583	2148	0.0071	5484.8	7659.8	174.70	22.66	31.83	282
270	0.811321	2194	0.0070	5600.5	7819.1	175.30	22.75	31.87	284
275	0.795720	2239	0.0068	5716.5	7978.6	175.88	22.83	31.92	287
280	0.780740	2284	0.0067	5832.8	8138.3	176.46	22.93	31.97	290
285	0.766342	2329	0.0066	5949.5	8298.3	177.02	23.02	32.03	292
290	0.752491	2374	0.0065	6066.6	8458.6	177.58	23.11	32.09	295
295	0.739156	2419	0.0063	6184.0	8619.2	178.13	23.20	32.15	297
300	0.726307	2464	0.0062	6301.8	8780.1	178.67	23.29	32.21	299

* TWO-PHASE BOUNDARY

TABLE 16. THERMODYNAMIC PROPERTIES OF FLUORINE

2.0 MN/m² ISOBAR

TEMPERATURE (IPTS 1968) K	DENSITY MOL/L	ISOTHERM DERIVATIVE J/MOL	ISOCHORE DERIVATIVE MN/m²-K	INTERNAL ENERGY J/MOL	ENTHALPY J/MOL	ENTROPY J/MOL-K	Cv J/MOL-K	Cp J/MOL-K	VELOCITY OF SOUND M/S
* 53.677	44.9188	27548	4.3192	-6069.0	-6024.4	60.87	36.44	54.45	1041
54	44.8665	27479	4.3189	-6051.3	-6006.7	61.20	36.38	54.59	1042
56	44.5403	26807	4.3167	-5941.4	-5896.5	63.22	36.03	55.66	1044
58	44.2171	26303	4.2336	-5831.3	-5786.1	65.16	35.67	55.88	1041
60	43.9022	26072	4.1432	-5722.2	-5676.7	67.02	35.34	55.83	1041
62	43.5794	25509	4.0548	-5611.9	-5566.0	68.84	35.05	56.09	1036
64	43.2484	24484	3.9605	-5500.5	-5454.2	70.62	34.67	56.60	1026
66	42.9171	23453	3.8441	-5388.8	-5342.2	72.34	34.24	56.82	1012
68	42.5990	22903	3.6726	-5279.0	-5232.1	73.99	33.77	55.84	998
70	42.2731	21809	3.5384	-5168.5	-5121.1	75.60	33.36	55.85	980
72	41.9291	20422	3.4224	-5055.6	-5007.9	77.20	32.98	56.47	959
74	41.5916	19925	3.3336	-4943.9	-4895.8	78.74	32.66	56.52	953
76	41.2489	19094	3.2127	-4831.4	-4783.0	80.25	32.24	56.38	937
78	40.9039	18031	3.0890	-4719.1	-4670.2	81.71	31.85	56.53	918
80	40.5536	17169	2.9883	-4606.1	-4556.8	83.15	31.52	56.82	903
82	40.1896	16209	2.8781	-4491.5	-4441.7	84.57	31.14	57.08	884
84	39.8395	15595	2.7694	-4379.0	-4328.8	85.93	30.80	56.82	870
86	39.4765	14702	2.6543	-4264.9	-4214.2	87.28	30.45	56.90	850
88	39.1079	13852	2.5397	-4150.2	-4099.0	88.60	30.11	56.90	830
90	38.7352	12988	2.4501	-4035.5	-3983.8	89.89	30.01	57.73	811
92	38.3532	12304	2.3658	-3919.5	-3867.3	91.16	29.82	58.27	795
94	37.9641	11668	2.2714	-3802.5	-3749.8	92.42	29.57	58.48	779
96	37.5703	10936	2.1812	-3685.1	-3631.9	93.65	29.32	58.91	760
98	37.1667	10304	2.0916	-3566.5	-3512.7	94.87	29.20	59.32	742
100	36.7560	9619	2.0053	-3446.9	-3392.5	96.07	28.98	59.93	723
102	36.3337	9025	1.9208	-3326.0	-3270.9	97.25	28.85	60.44	705
104	35.9010	8440	1.8359	-3203.4	-3147.7	98.44	28.67	60.90	687
106	35.4594	7793	1.7463	-3079.8	-3023.4	99.60	28.48	61.47	665
108	35.0039	7173	1.6605	-2954.3	-2897.1	100.76	28.34	62.22	644
110	34.5332	6586	1.5877	-2826.8	-2768.9	101.91	28.27	63.56	624
112	34.0428	6087	1.5060	-2696.5	-2637.8	103.06	28.13	64.14	604
114	33.5379	5511	1.4301	-2564.3	-2504.7	104.21	28.04	65.66	583
116	33.0082	4992	1.3500	-2428.7	-2368.1	105.36	27.94	66.81	560
118	32.4544	4442	1.2741	-2289.9	-2228.2	106.52	27.88	68.82	537
120	31.8680	3944	1.1969	-2146.7	-2083.9	107.68	27.82	70.74	514
122	31.2440	3484	1.1221	-1998.5	-1934.5	108.87	27.80	72.96	491
* 123.723	30.6702	3042	1.0561	-1866.1	-1800.9	109.91	27.80	76.04	468
* 123.723	2.62538	508	0.0282	2015.1	2776.9	146.90	26.08	54.12	167
124	2.61020	515	0.0279	2025.6	2791.8	147.02	25.98	53.58	167
126	2.50916	559	0.0263	2098.4	2895.5	147.85	25.39	50.23	171
128	2.42063	600	0.0250	2167.0	2993.2	148.62	24.89	47.64	174
130	2.34181	639	0.0239	2232.3	3086.4	149.34	24.47	45.58	177
132	2.27075	675	0.0228	2295.1	3175.8	150.03	24.12	43.91	180
134	2.20607	710	0.0220	2355.6	3262.2	150.68	23.82	42.52	183
136	2.14671	743	0.0212	2414.4	3346.0	151.30	23.55	41.34	185
138	2.09168	775	0.0204	2471.6	3427.7	151.89	23.33	40.34	188
140	2.04095	805	0.0198	2527.6	3507.5	152.47	23.13	39.48	190
142	1.99341	835	0.0192	2582.4	3585.7	153.02	22.95	38.73	193
144	1.94885	864	0.0187	2636.2	3662.5	153.56	22.80	38.07	195
146	1.90694	892	0.0181	2689.2	3738.0	154.08	22.66	37.49	197
148	1.86739	919	0.0177	2741.5	3812.5	154.59	22.55	36.97	199
150	1.82996	946	0.0172	2793.0	3886.4	155.08	22.44	36.51	201
152	1.79445	973	0.0168	2844.0	3958.6	155.56	22.35	36.10	203
154	1.76069	998	0.0164	2894.4	4030.4	156.03	22.27	35.72	205
156	1.72852	1024	0.0161	2944.4	4101.5	156.49	22.19	35.39	207
158	1.69781	1049	0.0157	2993.9	4171.9	156.94	22.13	35.08	209
160	1.66845	1073	0.0154	3043.1	4241.8	157.38	22.07	34.80	211
165	1.60027	1133	0.0147	3164.5	4414.3	158.44	21.96	34.21	216
170	1.53855	1192	0.0140	3284.2	4584.1	159.45	21.88	33.73	220
175	1.48229	1248	0.0134	3402.5	4751.7	160.43	21.83	33.34	224
180	1.43069	1304	0.0129	3519.7	4917.6	161.36	21.79	33.03	228
185	1.38312	1358	0.0124	3636.1	5082.1	162.26	21.78	32.77	232
190	1.33907	1411	0.0120	3751.8	5245.4	163.13	21.78	32.55	236
195	1.29811	1464	0.0116	3867.0	5407.7	163.98	21.80	32.38	239
200	1.25991	1515	0.0112	3981.8	5569.2	164.79	21.82	32.24	243
205	1.22415	1566	0.0108	4096.4	5730.1	165.59	21.85	32.13	246
210	1.19059	1617	0.0105	4210.7	5890.5	166.36	21.90	32.04	249
215	1.15901	1666	0.0102	4324.9	6050.5	167.11	21.94	31.97	253
220	1.12924	1716	0.0099	4439.1	6210.2	167.85	22.00	31.91	256
225	1.10109	1765	0.0097	4553.3	6369.7	168.56	22.06	31.88	259
230	1.07445	1813	0.0094	4667.6	6529.0	169.27	22.13	31.85	262
235	1.04916	1861	0.0092	4782.0	6688.2	169.95	22.20	31.84	265
240	1.02514	1909	0.0089	4896.5	6847.4	170.62	22.27	31.84	268
245	1.00228	1956	0.0087	5011.2	7006.6	171.28	22.35	31.85	271
250	0.980490	2003	0.0085	5126.1	7165.9	171.92	22.43	31.86	274
255	0.959694	2050	0.0083	5241.3	7325.3	172.55	22.51	31.89	276
260	0.939821	2107	0.0081	5356.7	7484.8	173.17	22.60	31.92	279
265	0.920809	2143	0.0080	5472.5	7644.5	173.78	22.68	31.95	282
270	0.902598	2189	0.0078	5588.5	7804.3	174.38	22.77	31.99	284
275	0.885136	2234	0.0076	5704.9	7964.4	174.96	22.86	32.04	287
280	0.868377	2280	0.0075	5821.5	8124.7	175.54	22.95	32.08	290
285	0.852276	2325	0.0073	5938.6	8285.2	176.11	23.04	32.14	292
290	0.836793	2371	0.0072	6056.0	8446.1	176.67	23.13	32.19	295
295	0.821892	2416	0.0071	6173.7	8607.1	177.22	23.23	32.25	297
300	0.807540	2461	0.0069	6291.9	8768.5	177.76	23.32	32.31	300

* TWO-PHASE BOUNDARY

TABLE 16. THERMODYNAMIC PROPERTIES OF FLUORINE

2.5 MN/m² ISOBAR

TEMPERATURE (IPTS 1968) K	DENSITY MOL/L	ISOTHERM DERIVATIVE J/MOL	ISOCHORE DERIVATIVE MN/m²-K	INTERNAL ENERGY J/MOL	ENTHALPY J/MOL	ENTROPY J/MOL-K	C_v J / MOL - K	C_p J / MOL - K	VELOCITY OF SOUND M/S
* 53.726	44.9289	27738	4.3213	-6079.4	-6023.8	60.89	36.42	54.34	1044
54	44.8846	27672	4.3210	-6064.5	-6008.8	61.17	36.37	54.46	1044
56	44.5589	26937	4.3186	-5954.9	-5898.7	63.17	36.03	55.55	1046
58	44.2360	26391	4.2408	-5845.0	-5788.5	65.12	35.66	55.86	1043
60	43.9213	26155	4.1487	-5736.1	-5679.1	66.98	35.34	55.81	1043
62	43.5990	25510	4.0549	-5625.9	-5568.6	68.79	35.08	56.11	1036
64	43.2689	24412	3.9587	-5514.8	-5457.0	70.57	34.72	56.66	1024
66	42.9385	23410	3.8479	-5403.4	-5345.1	72.30	34.29	56.93	1011
68	42.6208	22933	3.6802	-5293.7	-5235.0	73.95	33.79	55.90	999
70	42.2960	21866	3.5435	-5183.4	-5124.3	75.56	33.38	55.85	981
72	41.9536	20434	3.4251	-5070.9	-5011.4	77.15	33.01	56.50	959
74	41.6167	19883	3.3339	-4959.5	-4899.4	78.69	32.71	56.59	952
76	41.2751	19079	3.2166	-4847.2	-4786.7	80.20	32.28	56.48	937
78	40.9317	18053	3.0918	-4735.2	-4674.1	81.66	31.90	56.55	918
80	40.5827	17195	2.9906	-4622.6	-4561.0	83.10	31.57	56.84	903
82	40.2204	16199	2.8818	-4508.3	-4446.2	84.52	31.19	57.18	884
84	39.8715	15640	2.7744	-4396.1	-4333.4	85.87	30.84	56.85	871
86	39.5104	14763	2.6606	-4282.4	-4219.1	87.22	30.49	56.91	852
88	39.1440	13918	2.5480	-4168.1	-4104.2	88.54	30.14	56.93	832
90	38.7735	13093	2.4540	-4053.9	-3989.5	89.82	30.07	57.61	812
92	38.3937	12390	2.3687	-3938.4	-3873.2	91.10	29.90	58.17	796
94	38.0068	11739	2.2775	-3821.8	-3756.1	92.35	29.66	58.41	780
96	37.6159	11032	2.1862	-3705.0	-3638.5	93.58	29.39	58.79	762
98	37.2150	10397	2.0962	-3587.0	-3519.8	94.80	29.29	59.19	744
100	36.8077	9724	2.0099	-3468.0	-3400.1	95.99	29.07	59.74	725
102	36.3869	9121	1.9260	-3347.7	-3279.0	97.18	28.94	60.27	707
104	35.9599	8528	1.8422	-3225.9	-3156.4	98.35	28.77	60.77	689
106	35.5231	7908	1.7545	-3103.0	-3032.6	99.51	28.56	61.25	668
108	35.0730	7299	1.6704	-2978.4	-2907.1	100.66	28.40	61.96	647
110	34.6083	6721	1.5959	-2852.0	-2779.8	101.81	28.34	63.15	628
112	34.1242	6197	1.5168	-2722.7	-2649.5	102.96	28.20	63.91	608
114	33.6274	5658	1.4410	-2591.8	-2517.5	104.09	28.11	65.11	587
116	33.1069	5136	1.3632	-2457.6	-2382.1	105.24	27.99	66.29	566
118	32.5649	4608	1.2878	-2320.6	-2243.8	106.38	27.92	67.97	543
120	31.9922	4111	1.2116	-2179.4	-2101.2	107.54	27.85	69.71	520
122	31.3844	3638	1.1377	-2033.6	-1953.9	108.71	27.82	71.89	497
124	30.7376	3148	1.0644	-1882.7	-1801.3	109.90	27.83	75.06	473
126	30.0361	2694	0.9939	-1724.7	-1641.5	111.12	27.89	79.09	448
128	29.2661	2249	0.9069	-1558.0	-1472.6	112.39	27.91	82.57	418
* 128.076	29.2351	2231	0.9055	-1551.5	-1466.0	112.44	27.92	82.99	418
* 128.076	3.38107	438	0.0379	1971.5	2710.9	145.05	27.10	63.93	165
130	3.22908	490	0.0354	2053.9	2828.1	145.96	26.33	58.22	169
132	3.09563	539	0.0333	2132.5	2940.1	146.81	25.68	53.97	173
134	2.98026	583	0.0315	2205.8	3044.7	147.60	25.14	50.78	176
136	2.87857	625	0.0300	2275.1	3143.6	148.33	24.70	48.28	179
138	2.78763	664	0.0286	2341.3	3238.1	149.02	24.32	46.27	182
140	2.70537	701	0.0275	2404.9	3328.9	149.67	24.00	44.63	185
142	2.63030	736	0.0265	2466.3	3416.8	150.30	23.73	43.25	188
144	2.56126	770	0.0255	2526.0	3502.1	150.89	23.49	42.08	191
146	2.49737	802	0.0247	2584.2	3585.2	151.47	23.28	41.08	193
148	2.43794	834	0.0239	2641.0	3666.5	152.02	23.10	40.21	195
150	2.38240	865	0.0232	2696.7	3746.1	152.55	22.94	39.45	198
152	2.33029	894	0.0226	2751.5	3824.3	153.07	22.80	38.78	200
154	2.28124	923	0.0220	2805.4	3901.3	153.57	22.68	38.18	202
156	2.23491	951	0.0214	2858.5	3977.1	154.06	22.57	37.66	204
158	2.19104	979	0.0209	2910.9	4051.9	154.54	22.48	37.18	206
160	2.14940	1006	0.0204	2962.7	4125.9	155.01	22.40	36.76	208
165	2.05378	1072	0.0193	3090.0	4307.3	156.12	22.23	35.86	213
170	1.96839	1135	0.0184	3214.7	4484.8	157.18	22.11	35.15	218
175	1.89141	1196	0.0175	3337.3	4659.1	158.19	22.02	34.59	222
180	1.82147	1256	0.0168	3458.3	4830.8	159.16	21.97	34.12	227
185	1.75750	1313	0.0161	3578.0	5000.5	160.09	21.93	33.75	231
190	1.69867	1370	0.0155	3696.7	5168.4	160.99	21.92	33.44	235
195	1.64430	1425	0.0149	3814.5	5334.9	161.85	21.92	33.18	238
200	1.59384	1480	0.0144	3931.7	5500.3	162.69	21.93	32.97	242
205	1.54683	1533	0.0139	4048.4	5664.6	163.50	21.96	32.79	245
210	1.50289	1586	0.0135	4164.8	5828.2	164.29	21.99	32.65	249
215	1.46170	1638	0.0131	4280.8	5991.2	165.05	22.03	32.53	252
220	1.42298	1689	0.0127	4396.7	6153.6	165.80	22.09	32.44	256
225	1.38650	1739	0.0123	4512.5	6315.6	166.53	22.14	32.37	259
230	1.35204	1790	0.0120	4628.3	6477.3	167.24	22.20	32.31	262
235	1.31943	1839	0.0117	4744.1	6638.8	167.93	22.27	32.27	265
240	1.28851	1888	0.0114	4859.9	6800.1	168.61	22.34	32.25	268
245	1.25915	1937	0.0111	4975.8	6961.3	169.28	22.42	32.23	271
250	1.23121	1986	0.0108	5091.9	7122.5	169.93	22.49	32.23	274
255	1.20459	2034	0.0106	5208.2	7283.6	170.57	22.58	32.23	276
260	1.17920	2081	0.0103	5324.7	7444.8	171.19	22.66	32.25	279
265	1.15493	2129	0.0101	5441.5	7606.1	171.81	22.74	32.26	282
270	1.13173	2176	0.0099	5558.5	7767.5	172.41	22.83	32.29	285
275	1.10950	2223	0.0097	5675.7	7929.0	173.00	22.92	32.32	287
280	1.08819	2269	0.0095	5793.3	8090.7	173.59	23.01	32.36	290
285	1.06775	2316	0.0093	5911.2	8252.6	174.16	23.10	32.40	292
290	1.04810	2362	0.0091	6029.4	8414.7	174.72	23.19	32.44	295
295	1.02922	2408	0.0089	6148.0	8577.0	175.28	23.28	32.49	297
300	1.01104	2454	0.0088	6266.9	8739.6	175.83	23.38	32.54	300

* TWO-PHASE BOUNDARY

TABLE 16. THERMODYNAMIC PROPERTIES OF FLUORINE

ISOTHERM DERIVATIVE J/MOL	ISOCHORE DERIVATIVE MN/m²-K	INTERNAL ENERGY J/MOL	ENTHALPY J/MOL	ENTROPY J/MOL-K	C_v J/MOL-K	C_p J/MOL-K	VELOCITY OF SOUND M/S
27925	4.3233	-6089.9	-6023.1	60.90	36.41	54.23	1046
27863	4.3230	-6077.6	-6010.8	61.13	36.37	54.33	1047
27066	4.3204	-5968.3	-5901.0	63.13	36.02	55.45	1047
26479	4.2479	-5858.6	-5790.8	65.08	35.65	55.83	1045
26238	4.1547	-5749.9	-5681.6	66.93	35.34	55.78	1044
25516	4.0556	-5640.0	-5571.2	68.75	35.11	56.12	1036
24350	3.9572	-5529.0	-5459.7	70.53	34.76	56.72	1023
23374	3.8513	-5417.9	-5348.0	72.25	34.33	57.02	1011
22965	3.6874	-5308.3	-5238.0	73.90	33.82	55.96	1000
21925	3.5487	-5198.3	-5127.4	75.51	33.40	55.85	982
20453	3.4282	-5086.2	-5014.8	77.11	33.04	56.51	960
19850	3.3347	-4975.0	-4902.9	78.64	32.79	56.66	951
19070	3.2203	-4863.0	-4790.4	80.15	32.33	56.56	937
18080	3.0948	-4751.3	-4678.1	81.61	31.94	56.57	918
17225	2.9931	-4639.1	-4565.2	83.04	31.62	56.85	903
16197	2.8853	-4525.1	-4450.6	84.46	31.24	57.25	884
15688	2.7793	-4413.2	-4338.0	85.82	30.89	56.87	872
14827	2.6669	-4299.9	-4224.0	87.16	30.53	56.92	853
13986	2.5559	-4186.0	-4109.4	88.48	30.17	56.94	834
13198	2.4585	-4072.3	-3995.0	89.76	30.13	57.49	814
12478	2.3720	-3957.2	-3879.1	91.03	29.98	58.06	798
11813	2.2808	-3841.1	-3762.3	92.28	29.74	58.33	781
11129	2.1911	-3724.8	-3645.1	93.51	29.46	58.66	764
10491	2.1012	-3607.3	-3526.8	94.72	29.38	59.08	745
9829	2.0144	-3489.0	-3407.6	95.92	29.16	59.55	727
9219	1.9315	-3369.3	-3287.0	97.10	29.03	60.11	709
8620	1.8483	-3248.2	-3164.9	98.27	28.86	60.63	690
8024	1.7630	-3126.0	-3041.7	99.43	28.63	61.05	671
7425	1.6801	-3002.3	-2916.9	100.57	28.46	61.70	651
6852	1.6043	-2876.9	-2790.4	101.71	28.41	62.76	631
6310	1.5273	-2748.6	-2660.9	102.85	28.26	63.64	612
5799	1.4517	-2619.0	-2530.0	103.98	28.16	64.61	592
5277	1.3758	-2486.1	-2395.7	105.12	28.04	65.78	571
4766	1.3010	-2350.6	-2258.8	106.25	27.96	67.22	549
4270	1.2265	-2211.3	-2117.9	107.40	27.87	68.87	527
3790	1.1527	-2067.7	-1972.5	108.55	27.85	70.90	504
3317	1.0810	-1919.6	-1822.5	109.72	27.84	73.62	480
2864	1.0099	-1765.3	-1666.0	110.92	27.97	77.11	456
2425	0.9286	-1602.8	-1501.0	112.16	27.85	80.21	429
1962	0.8520	-1431.1	-1326.4	113.45	28.01	86.56	399
1554	0.7776	-1260.7	-1152.9	114.71	28.13	94.42	371
364	0.0495	1905.5	2616.0	143.30	28.17	78.05	163
369	0.0491	1914.4	2628.6	143.39	28.07	76.97	163
434	0.0451	2015.3	2771.7	144.47	27.05	66.96	168
489	0.0419	2104.6	2898.8	145.41	26.27	60.50	172
540	0.0394	2186.1	3015.0	146.26	25.63	55.93	176
586	0.0373	2262.0	3123.2	147.04	25.12	52.50	180
629	0.0355	2333.6	3225.5	147.76	24.69	49.84	183
669	0.0339	2401.7	3322.9	148.45	24.32	47.70	186
708	0.0326	2467.1	3416.5	149.09	24.01	45.94	189
744	0.0314	2530.3	3506.9	149.71	23.75	44.47	192
779	0.0303	2591.6	3594.5	150.29	23.52	43.22	194
813	0.0293	2651.2	3679.9	150.86	23.32	42.15	197
846	0.0284	2709.4	3763.2	151.40	23.15	41.23	199
877	0.0275	2766.5	3844.9	151.93	23.00	40.42	201
908	0.0268	2822.5	3925.0	152.44	22.86	39.70	204
938	0.0261	2877.6	4003.7	152.94	22.75	39.07	206
1010	0.0245	3011.9	4195.7	154.12	22.51	37.77	211
1079	0.0231	3142.3	4381.9	155.23	22.34	36.76	216
1144	0.0220	3269.8	4563.6	156.28	22.22	35.96	221
1208	0.0210	3395.0	4741.8	157.29	22.14	35.32	225
1269	0.0200	3518.3	4917.1	158.25	22.09	34.81	229
1329	0.0192	3640.2	5090.0	159.17	22.05	34.38	233
1387	0.0185	3760.9	5261.0	160.06	22.04	34.03	237
1444	0.0178	3880.7	5430.4	160.92	22.04	33.74	241
1500	0.0172	3999.7	5598.4	161.75	22.06	33.49	245
1555	0.0166	4118.2	5765.4	162.55	22.09	33.29	248
1609	0.0161	4236.2	5931.4	163.33	22.13	33.13	252
1663	0.0156	4353.9	6096.7	164.09	22.17	32.99	255
1715	0.0151	4471.4	6261.3	164.83	22.22	32.88	258
1767	0.0147	4588.6	6425.5	165.55	22.28	32.79	262
1818	0.0143	4705.9	6589.3	166.26	22.34	32.72	265
1869	0.0139	4823.0	6752.7	166.95	22.41	32.67	268
1919	0.0135	4940.3	6915.9	167.62	22.48	32.63	271
1969	0.0132	5057.6	7079.0	168.28	22.56	32.60	274
2018	0.0129	5175.0	7242.0	168.92	22.64	32.58	276
2067	0.0126	5292.6	7404.9	169.56	22.72	32.58	279
2116	0.0123	5410.3	7567.7	170.18	22.81	32.58	282
2164	0.0120	5528.3	7730.7	170.79	22.89	32.59	285
2212	0.0117	5646.5	7893.7	171.38	22.98	32.61	287
2259	0.0115	5765.0	8056.8	171.97	23.07	32.63	290
2307	0.0113	5883.7	8220.0	172.55	23.16	32.66	293
2354	0.0110	6002.8	8383.4	173.12	23.25	32.69	295
2401	0.0108	6122.1	8547.0	173.68	23.34	32.73	298
2447	0.0106	6241.8	8710.7	174.23	23.43	32.77	300

TABLE 16. THERMODYNAMIC PROPERTIES OF FLUORINE

3.5 MN/m² ISOBAR

TEMPERATURE (IPTS 1968) K	DENSITY MOL/L	ISOTHERM DERIVATIVE J/MOL	ISOCHORE DERIVATIVE MN/m²-K	INTERNAL ENERGY J/MOL	ENTHALPY J/MOL	ENTROPY J/MOL-K	C_v J / MOL - K	C_p	VELOCITY OF SOUND M/S
* 53.823	44.9490	28107	4.3252	-6100.4	-6022.5	60.91	36.39	54.13	1049
54	44.9205	28053	4.3250	-6090.8	-6012.9	61.09	36.36	54.20	1049
56	44.5959	27195	4.3222	-5981.7	-5903.2	63.09	36.01	55.35	1049
58	44.2738	26566	4.2549	-5872.2	-5793.1	65.03	35.64	55.81	1046
60	43.9594	26321	4.1613	-5763.7	-5684.0	66.89	35.34	55.76	1046
62	43.6382	25527	4.0570	-5654.0	-5573.8	68.71	35.14	56.13	1036
64	43.3100	24298	3.9558	-5543.3	-5462.5	70.49	34.80	56.78	1021
66	42.9813	23346	3.8543	-5432.4	-5351.0	72.21	34.38	57.11	1010
68	42.6644	22999	3.6940	-5323.0	-5240.9	73.86	33.84	56.01	1001
70	42.3416	21986	3.5539	-5213.2	-5130.5	75.47	33.42	55.85	983
72	42.0025	20478	3.4319	-5101.5	-5018.1	77.06	33.06	56.53	960
74	41.6671	19827	3.3358	-4990.5	-4906.6	78.59	32.80	56.72	950
76	41.3275	19067	3.2238	-4878.8	-4794.1	80.10	32.38	56.53	937
78	40.9870	18111	3.0979	-4767.4	-4682.0	81.56	31.98	56.58	918
80	40.6408	17260	2.9957	-4655.5	-4569.3	82.99	31.67	56.86	903
82	40.2822	16204	2.8888	-4541.9	-4455.1	84.41	31.29	57.32	884
84	39.9353	15739	2.7840	-4430.3	-4342.6	85.76	30.94	56.88	873
86	39.5779	14894	2.6730	-4317.3	-4228.9	87.10	30.58	56.92	854
88	39.2154	14058	2.5635	-4203.8	-4114.5	88.42	30.20	56.95	835
90	38.8493	13303	2.4636	-4090.6	-4000.5	89.70	30.18	57.39	816
92	38.4738	12568	2.3756	-3975.9	-3884.9	90.97	30.05	57.96	799
94	38.0915	11890	2.2846	-3860.3	-3768.4	92.22	29.82	58.26	782
96	37.7057	11227	2.1958	-3744.5	-3651.7	93.44	29.53	58.53	765
98	37.3103	10587	2.1065	-3627.6	-3533.7	94.65	29.46	58.96	747
100	36.9095	9935	2.0190	-3509.8	-3415.0	95.84	29.24	59.36	729
102	36.4973	9318	1.9373	-3390.8	-3294.9	97.02	29.11	59.95	711
104	36.0760	8714	1.8545	-3270.3	-3173.3	98.19	28.95	60.49	692
106	35.6477	8140	1.7712	-3148.9	-3050.7	99.34	28.70	60.85	674
108	35.2077	7550	1.6893	-3026.0	-2926.6	100.48	28.51	61.45	654
110	34.7543	6982	1.6128	-2901.5	-2800.8	101.61	28.48	62.40	635
112	34.2827	6428	1.5374	-2774.3	-2672.2	102.75	28.32	63.36	615
114	33.7999	5937	1.4621	-2645.7	-2542.1	103.87	28.22	64.15	596
116	33.2965	5415	1.3879	-2514.1	-2409.0	105.00	28.08	65.30	576
118	32.7748	4918	1.3138	-2380.2	-2273.4	106.13	27.99	66.55	555
120	32.2265	4425	1.2408	-2242.6	-2134.0	107.26	27.90	68.10	533
122	31.6484	3940	1.1681	-2101.1	-1990.5	108.40	27.87	70.05	511
124	31.0394	3480	1.0971	-1955.5	-1842.8	109.56	27.86	72.37	486
126	30.3858	3028	1.0259	-1804.3	-1689.1	110.74	28.01	75.43	463
128	29.6793	2595	0.9512	-1645.6	-1527.7	111.95	27.83	78.50	439
130	28.9100	2150	0.8753	-1479.5	-1358.4	113.20	27.98	83.41	411
132	28.0436	1720	0.7956	-1301.1	-1176.3	114.52	27.91	89.67	381
134	27.0363	1290	0.7162	-1106.3	-976.8	115.95	28.24	101.14	349
* 135.161	26.3502	1046	0.6645	-981.3	-848.5	116.86	28.27	110.45	328
* 135.161	5.18934	285	0.0638	1813.5	2487.9	141.54	29.36	101.00	161
136	5.01713	321	0.0606	1870.4	2568.0	142.13	28.71	90.62	163
138	4.69405	394	0.0549	1986.9	2732.5	143.33	27.52	75.44	168
140	4.44564	455	0.0506	2086.5	2873.8	144.35	26.63	66.52	173
142	4.24282	510	0.0473	2175.6	3000.6	145.25	25.93	60.56	177
144	4.07110	560	0.0446	2257.4	3117.1	146.06	25.37	56.24	181
146	3.92204	606	0.0423	2333.8	3226.2	146.81	24.91	52.96	184
148	3.79033	649	0.0404	2406.0	3329.4	147.52	24.52	50.37	187
150	3.67234	690	0.0386	2474.9	3428.0	148.18	24.19	48.28	190
152	3.56549	728	0.0371	2541.1	3522.8	148.81	23.91	46.54	193
154	3.46789	765	0.0358	2605.1	3614.4	149.40	23.67	45.09	196
156	3.37809	801	0.0346	2667.2	3703.3	149.98	23.47	43.85	198
158	3.29497	835	0.0334	2727.7	3789.9	150.53	23.29	42.78	201
160	3.21764	868	0.0324	2786.7	3874.5	151.06	23.13	41.85	203
165	3.04517	947	0.0302	2929.5	4078.8	152.32	22.82	39.98	209
170	2.89638	1022	0.0284	3066.7	4275.1	153.49	22.59	38.58	214
175	2.76585	1093	0.0268	3199.7	4465.2	154.59	22.43	37.50	219
180	2.64983	1160	0.0255	3329.6	4650.4	155.64	22.32	36.64	224
185	2.54565	1226	0.0243	3457.0	4831.9	156.63	22.24	35.95	228
190	2.45130	1289	0.0232	3582.4	5010.2	157.58	22.19	35.39	233
195	2.36524	1350	0.0222	3706.2	5185.9	158.50	22.17	34.93	237
200	2.28626	1410	0.0214	3828.7	5359.6	159.38	22.16	34.55	241
205	2.21341	1468	0.0206	3950.2	5531.5	160.22	22.17	34.23	244
210	2.14591	1526	0.0198	4071.0	5702.0	161.05	22.19	33.96	248
215	2.08311	1582	0.0192	4191.0	5871.2	181.84	22.22	33.74	251
220	2.02448	1637	0.0186	4310.6	6039.4	162.62	22.25	33.56	255
225	1.96956	1691	0.0180	4429.8	6206.8	163.37	22.30	33.40	258
230	1.91798	1745	0.0175	4548.7	6373.5	164.12	22.36	33.28	261
235	1.86941	1798	0.0170	4667.4	6539.6	164.82	22.42	33.17	265
240	1.82357	1850	0.0165	4786.0	6705.3	165.51	22.48	33.09	268
245	1.78020	1902	0.0161	4904.5	6870.6	166.19	22.55	33.02	271
250	1.73909	1953	0.0156	5023.0	7035.5	166.86	22.62	32.97	274
255	1.70006	2003	0.0152	5141.6	7200.3	167.51	22.70	32.94	277
260	1.66294	2054	0.0149	5260.2	7364.9	168.15	22.78	32.91	279
265	1.62758	2103	0.0145	5379.0	7529.5	168.78	22.86	32.90	282
270	1.59384	2153	0.0142	5498.0	7694.0	169.39	22.95	32.89	285
275	1.56161	2201	0.0139	5617.1	7858.4	170.00	23.04	32.90	288
280	1.53077	2250	0.0136	5736.5	8022.9	170.59	23.12	32.91	290
285	1.50125	2298	0.0133	5856.1	8187.5	171.17	23.21	32.92	293
290	1.47294	2346	0.0130	5976.0	8352.2	171.75	23.30	32.95	295
295	1.44577	2394	0.0127	6096.1	8517.0	172.31	23.39	32.97	298
300	1.41967	2441	0.0125	6216.6	8681.9	172.86	23.49	33.01	300

* TWO-PHASE BOUNDARY

TABLE 16. THERMODYNAMIC PROPERTIES OF FLUORINE

2 ISOBAR

DENSITY MOL/L	ISOTHERM DERIVATIVE J/MOL	ISOCHORE DERIVATIVE MN/m²-K	INTERNAL ENERGY J/MOL	ENTHALPY J/MOL	ENTROPY J/MOL-K	C_v J / MOL - K	C_p J / MOL - K	VELOCITY OF SOUND M/S
4.9589	28285	4.3270	-6110.9	-6021.9	60.92	36.37	54.02	1051
4.9383	28241	4.3269	-6103.9	-6014.9	61.05	36.35	54.08	1052
4.6142	27323	4.3239	-5995.0	-5905.4	63.05	36.00	55.26	1051
4.2926	26653	4.2619	-5885.8	-5795.5	64.99	35.64	55.78	1048
3.9784	26403	4.1684	-5777.4	-5686.5	66.85	35.33	55.75	1047
3.6678	25544	4.0591	-5667.9	-5576.3	68.67	35.16	56.15	1036
3.3306	24255	3.9545	-5557.5	-5465.2	70.44	34.84	56.82	1020
3.0027	23326	3.8563	-5446.9	-5353.9	72.16	34.43	57.18	1010
2.6861	23036	3.7001	-5337.6	-5243.9	73.82	33.87	56.05	1002
2.3643	22049	3.5591	-5228.1	-5133.6	75.42	33.44	55.85	984
2.0269	20509	3.4373	-5116.7	-5021.5	77.01	33.08	56.57	961
1.6923	19813	3.3373	-5006.0	-4910.1	78.55	32.84	56.77	949
1.3537	19071	3.2271	-4894.5	-4797.8	80.05	32.43	56.69	937
1.0145	18147	3.1021	-4783.4	-4685.9	81.51	32.02	56.60	919
0.6697	17300	2.9985	-4671.8	-4573.5	82.94	31.72	56.86	903
0.3130	16219	2.8921	-4558.7	-4459.5	84.35	31.35	57.37	884
9.9670	15792	2.7885	-4447.3	-4347.2	85.71	30.99	56.88	873
9.6113	14962	2.6790	-4334.7	-4233.7	87.05	30.62	56.91	855
9.2509	14134	2.5708	-4221.5	-4119.6	88.36	30.23	56.94	837
8.8867	13407	2.4692	-4108.8	-4005.9	89.64	30.23	57.29	818
8.5134	12660	2.3795	-3994.6	-3890.7	90.90	30.11	57.85	800
8.1334	11971	2.2887	-3879.4	-3774.5	92.15	29.90	58.18	783
7.7501	11326	2.2006	-3764.1	-3658.1	93.37	29.60	58.40	767
7.3574	10684	2.1121	-3647.7	-3540.6	94.58	29.53	58.85	749
6.9595	10042	2.0237	-3530.5	-3422.3	95.77	29.31	59.17	730
6.5507	9419	1.9433	-3412.1	-3302.7	96.94	29.18	59.79	713
6.1330	8811	1.8610	-3292.3	-3181.6	98.11	29.03	60.35	694
5.7087	8256	1.7790	-3171.6	-3059.5	99.25	28.77	60.63	677
5.2734	7674	1.6983	-3049.5	-2936.1	100.39	28.57	61.19	658
4.8252	7112	1.6214	-2925.9	-2811.0	101.52	28.53	62.06	638
4.3598	6548	1.5472	-2799.7	-2683.3	102.65	28.38	63.06	619
3.8832	6070	1.4723	-2672.1	-2554.0	103.77	28.27	63.73	600
3.3877	5551	1.3995	-2541.8	-2422.0	104.89	28.13	64.85	580
2.8750	5065	1.3262	-2409.2	-2287.5	106.00	28.03	65.94	560
2.3376	4574	1.2544	-2273.2	-2149.5	107.13	27.92	67.40	539
1.7730	4086	1.1829	-2133.7	-2007.8	108.26	27.88	69.25	517
1.1799	3638	1.1128	-1990.4	-1862.1	109.40	27.88	71.29	495
0.5466	3188	1.0422	-1842.0	-1711.1	110.56	28.03	74.03	471
9.8661	2759	0.9713	-1686.7	-1552.8	111.75	27.89	76.92	448
9.1333	2328	0.8972	-1525.3	-1386.0	112.96	27.92	80.88	421
8.3192	1910	0.8218	-1353.3	-1212.0	114.24	27.77	85.97	394
7.3957	1495	0.7446	-1168.5	-1022.5	115.59	28.13	94.34	363
6.2965	1087	0.6603	-951.0	-798.9	117.08	28.16	107.04	330
4.8788	649	0.5636	-722.7	-561.9	118.81	29.79	138.90	282
4.7511	614	0.5539	-703.4	-541.8	118.96	30.05	142.67	277
6.35753	205	0.0806	1684.9	2314.1	139.63	31.10	139.49	156
5.78681	294	0.0715	1844.3	2535.5	141.22	28.98	101.69	165
5.37988	372	0.0642	1973.4	2716.9	142.51	27.70	82.04	170
5.07587	438	0.0590	2081.2	2869.3	143.58	26.77	71.18	175
4.83178	495	0.0549	2176.3	3004.2	144.51	26.05	64.14	179
4.62739	547	0.0516	2242.8	3127.2	145.34	25.47	59.16	183
4.45142	595	0.0489	2343.0	3241.6	146.11	25.00	55.42	186
4.29688	640	0.0466	2418.5	3349.4	146.82	24.61	52.50	190
4.15913	682	0.0445	2490.2	3452.0	147.49	24.28	50.15	193
4.03491	723	0.0427	2558.9	3550.3	148.13	24.00	48.22	195
3.92182	761	0.0411	2625.2	3645.1	148.73	23.76	46.61	198
3.81808	798	0.0397	2689.3	3736.9	149.31	23.55	45.24	201
3.59123	884	0.0367	2842.3	3956.1	150.66	23.14	42.58	207
3.39981	965	0.0342	2987.4	4164.0	151.90	22.86	40.66	213
3.23465	1041	0.0321	3126.9	4363.5	153.06	22.65	39.21	218
3.08977	1113	0.0303	3262.0	4556.6	154.15	22.50	38.09	223
2.96102	1182	0.0288	3393.9	4744.8	155.18	22.40	37.20	227
2.84541	1249	0.0274	3523.1	4928.9	156.16	22.34	36.48	232
2.74072	1313	0.0262	3650.3	5109.7	157.10	22.29	35.89	236
2.64524	1376	0.0251	3775.8	5287.9	158.00	22.27	35.40	240
2.55762	1437	0.0242	3900.4	5463.9	158.87	22.27	35.00	244
2.47681	1497	0.0233	4023.1	5638.0	159.71	22.28	34.66	248
2.40192	1555	0.0224	4145.3	5810.6	160.52	22.31	34.38	251
2.33226	1612	0.0217	4266.8	5981.9	161.31	22.34	34.14	255
2.26721	1668	0.0210	4387.8	6152.1	162.07	22.38	33.94	258
2.20629	1724	0.0203	4508.4	6321.4	162.82	22.43	33.78	261
2.14906	1778	0.0197	4628.6	6489.9	163.54	22.49	33.64	265
2.09517	1832	0.0192	4748.6	6657.8	164.25	22.55	33.52	268
2.04429	1885	0.0186	4868.5	6825.2	164.94	22.62	33.43	271
1.99616	1937	0.0182	4988.3	6992.1	165.61	22.69	33.36	274
1.95053	1989	0.0177	5108.0	7158.7	166.27	22.76	33.30	277
1.90720	2041	0.0172	5227.8	7325.1	166.92	22.84	33.25	280
1.86598	2091	0.0168	5347.6	7491.3	167.55	22.92	33.22	282
1.82671	2142	0.0164	5467.6	7657.3	168.17	23.00	33.20	285
1.78924	2192	0.0161	5587.7	7823.3	168.78	23.09	33.19	288
1.75343	2241	0.0157	5708.0	7989.2	159.38	23.18	33.18	291
1.71918	2290	0.0154	5828.4	8155.1	169.97	23.26	33.19	293
1.68637	2339	0.0150	5949.1	8321.1	170.55	23.35	33.20	296
1.65491	2388	0.0147	6070.1	8487.1	171.11	23.44	33.22	298
1.62471	2436	0.0144	6191.3	8653.2	171.67	23.54	33.24	301

BOUNDARY

TABLE 16. THERMODYNAMIC PROPERTIES OF FLUORINE

4.5 MN/m² ISOBAR

TEMPERATURE (IPTS 1968) K	DENSITY MOL/L	ISOTHERM DERIVATIVE J/MOL	ISOCHORE DERIVATIVE MN/m²-K	INTERNAL ENERGY J/MOL	ENTHALPY J/MOL	ENTROPY J/MOL-K	C_v J / MOL - K	C_p	VELOCITY OF SOUND M/S
* 53.920	44.9688	28458	4.3288	-6121.3	-6021.2	60.93	36.36	53.92	1054
54	44.9559	28429	4.3287	-6117.0	-6016.9	61.01	36.35	53.96	1054
56	44.6325	27451	4.3256	-6008.4	-5907.6	63.01	36.00	55.16	1052
58	44.3113	26739	4.2687	-5899.3	-5797.8	64.95	35.63	55.76	1049
60	43.9973	26485	4.1761	-5791.1	-5688.9	66.81	35.33	55.74	1049
62	43.6774	25566	4.0618	-5681.9	-5578.9	68.63	35.19	56.16	1036
64	43.3512	24222	3.9534	-5571.8	-5468.0	70.40	34.89	56.86	1019
66	43.0241	23313	3.8545	-5461.4	-5356.8	72.12	34.47	57.20	1009
68	42.7078	23075	3.7058	-5352.1	-5246.8	73.77	33.90	56.08	1002
70	42.3870	22114	3.5643	-5242.9	-5136.7	75.38	33.46	55.84	986
72	42.0513	20546	3.4428	-5131.9	-5024.9	76.96	33.11	56.59	961
74	41.7175	19808	3.3393	-5021.5	-4913.6	78.50	32.88	56.82	949
76	41.3799	19081	3.2301	-4910.2	-4801.5	80.00	32.47	56.74	937
78	41.0421	18188	3.1071	-4799.4	-4689.8	81.46	32.06	56.64	920
80	40.6986	17344	3.0014	-4688.2	-4577.6	82.89	31.77	56.86	904
82	40.3438	16242	2.8954	-4575.4	-4463.9	84.30	31.40	57.40	884
84	39.9986	15848	2.7930	-4464.3	-4351.8	85.65	31.04	56.88	874
86	39.6447	15034	2.6847	-4352.0	-4238.5	86.99	30.67	56.90	857
88	39.2862	14212	2.5778	-4239.2	-4124.7	88.30	30.27	56.93	839
90	38.9239	13511	2.4753	-4126.9	-4011.3	89.58	30.26	57.20	820
92	38.5528	12754	2.3840	-4013.1	-3896.4	90.84	30.18	57.76	802
94	38.1750	12055	2.2932	-3898.5	-3780.6	92.09	29.97	58.10	784
96	37.7940	11428	2.2053	-3783.6	-3664.5	93.31	29.66	58.26	769
98	37.4039	10783	2.1179	-3667.7	-3547.4	94.51	29.60	58.74	750
100	37.0090	10150	2.0287	-3551.0	-3429.4	95.69	29.39	58.99	732
102	36.6035	9522	1.9493	-3433.3	-3310.3	96.86	29.24	59.62	715
104	36.1894	8910	1.8678	-3314.1	-3189.8	98.03	29.11	60.20	696
106	35.7688	8371	1.7867	-3194.0	-3068.2	99.17	28.83	60.42	680
108	35.3380	7798	1.7071	-3072.7	-2945.4	100.30	28.62	60.94	661
110	34.8949	7241	1.6302	-2950.0	-2821.1	101.43	28.58	61.74	642
112	34.4354	6670	1.5567	-2824.8	-2694.1	102.55	28.43	62.75	622
114	33.9647	6200	1.4824	-2698.2	-2565.7	103.66	28.31	63.34	604
116	33.4767	5685	1.4107	-2569.0	-2434.6	104.77	28.17	64.41	585
118	32.9724	5207	1.3382	-2437.7	-2301.2	105.88	28.06	65.39	565
120	32.4452	4719	1.2675	-2303.3	-2164.6	107.00	27.95	66.75	545
122	31.8932	4234	1.1971	-2165.6	-2024.5	108.12	27.90	68.49	523
124	31.3145	3791	1.1278	-2024.4	-1880.7	109.24	27.89	70.31	502
126	30.6997	3344	1.0588	-1878.5	-1732.0	110.38	28.04	72.86	478
128	30.0423	2918	0.9891	-1726.3	-1576.5	111.56	27.89	75.44	456
130	29.3405	2499	0.9172	-1568.9	-1415.5	112.74	27.88	78.71	431
132	28.5692	2090	0.8449	-1402.2	-1244.7	113.98	27.71	82.94	406
134	27.7102	1687	0.7720	-1225.2	-1062.8	115.28	27.95	89.62	377
136	26.7183	1288	0.6934	-1003.7	-835.2	116.68	28.03	99.14	346
138	25.5267	899	0.6118	-800.0	-623.8	118.22	28.53	116.66	311
140	23.8951	487	0.5134	-547.7	-359.4	120.12	30.51	163.28	262
* 140.855	22.8479	305	0.4583	-401.5	-204.6	121.22	30.93	216.95	237
* 140.855	7.87301	123	0.1055	1501.1	2072.6	137.39	34.06	240.05	151
142	7.14432	201	0.0927	1664.6	2294.5	138.96	31.32	149.90	159
144	6.44627	297	0.0800	1844.6	2542.7	140.70	28.95	103.66	167
146	5.99161	370	0.0725	1978.2	2729.2	141.98	27.62	85.38	174
148	5.64689	436	0.0665	2090.6	2887.5	143.06	26.71	73.82	178
150	5.37088	494	0.0619	2189.3	3027.2	144.00	26.00	66.34	182
152	5.14031	547	0.0582	2278.9	3154.3	144.84	25.44	61.04	186
154	4.94221	596	0.0551	2361.7	3272.2	145.61	24.99	57.08	189
156	4.76855	642	0.0524	2439.5	3383.2	146.33	24.61	53.99	193
158	4.61398	685	0.0501	2513.3	3488.6	147.00	24.29	51.51	196
160	4.47479	726	0.0481	2583.9	3589.5	147.63	24.02	49.47	198
165	4.17777	821	0.0439	2749.6	3826.7	149.09	23.50	45.67	205
170	3.93371	909	0.0406	2904.2	4048.2	150.42	23.13	43.05	211
175	3.72715	990	0.0379	3051.0	4258.4	151.63	22.88	41.14	216
180	3.54857	1067	0.0356	3192.2	4460.3	152.77	22.69	39.68	222
185	3.39168	1140	0.0337	3329.0	4655.8	153.84	22.57	38.55	226
190	3.25211	1210	0.0320	3462.4	4846.1	154.86	22.48	37.64	231
195	3.12668	1278	0.0305	3593.2	5032.4	155.83	22.42	36.91	235
200	3.01301	1343	0.0291	3721.9	5215.4	156.75	22.39	36.30	239
205	2.90928	1407	0.0279	3848.9	5395.7	157.64	22.38	35.81	243
210	2.81406	1469	0.0268	3974.5	5573.6	158.50	22.38	35.39	247
215	2.72618	1529	0.0259	4099.0	5749.7	159.33	22.40	35.04	251
220	2.64472	1588	0.0249	4222.6	5924.1	160.13	22.42	34.74	254
225	2.56890	1646	0.0241	4345.5	6097.2	160.91	22.46	34.50	258
230	2.49809	1703	0.0233	4467.7	6269.1	161.67	22.51	34.29	261
235	2.43175	1759	0.0226	4589.6	6440.1	162.40	22.56	34.11	265
240	2.36941	1815	0.0219	4711.1	6610.3	163.12	22.62	33.96	268
245	2.31068	1869	0.0213	4832.3	6779.8	163.82	22.68	33.84	271
250	2.25522	1923	0.0207	4953.4	6948.7	164.50	22.75	33.74	274
255	2.20274	1976	0.0202	5074.3	7117.2	165.17	22.82	33.66	277
260	2.15297	2028	0.0197	5195.2	7285.4	165.82	22.90	33.59	280
265	2.10570	2080	0.0192	5316.1	7453.2	166.46	22.98	33.54	283
270	2.06072	2132	0.0187	5437.1	7620.8	167.09	23.06	33.50	286
275	2.01785	2183	0.0183	5558.1	7788.2	167.70	23.14	33.48	288
280	1.97694	2233	0.0179	5679.3	7955.6	168.30	23.23	33.46	291
285	1.93783	2283	0.0175	5800.7	8122.8	168.89	23.31	33.45	294
290	1.90041	2333	0.0171	5922.2	8290.1	169.48	23.40	33.45	296
295	1.86456	2382	0.0167	6043.9	8457.4	170.05	23.49	33.46	299
300	1.83017	2431	0.0164	6165.9	8624.7	170.61	23.58	33.47	301

* TWO-PHASE BOUNDARY

TABLE 16. THERMODYNAMIC PROPERTIES OF FLUORINE

/m² ISOBAR

DENSITY MOL/L	ISOTHERM DERIVATIVE J/MOL	ISOCHORE DERIVATIVE MN/m²-K	INTERNAL ENERGY J/MOL	ENTHALPY J/MOL	ENTROPY J/MOL-K	C_v J / MOL - K	C_p J / MOL - K	VELOCITY OF SOUND M/S
44.9785	28627	4.3305	-6131.8	-6020.6	60.94	36.35	53.82	1056
44.9734	28615	4.3305	-6130.0	-6018.9	60.98	36.34	53.84	1056
44.5506	27577	4.3272	-6021.7	-5909.7	62.98	35.99	55.06	1054
44.3300	26826	4.2755	-5912.9	-5800.1	64.91	35.62	55.74	1051
44.0162	26567	4.1830	-5804.9	-5691.3	66.77	35.32	55.72	1050
43.6969	25594	4.0651	-5695.9	-5581.5	68.58	35.21	56.18	1037
43.3718	24199	3.9524	-5586.0	-5470.7	70.35	34.93	56.89	1019
43.0456	23307	3.8529	-5475.8	-5359.7	72.08	34.52	57.21	1008
42.7294	23117	3.7110	-5366.7	-5249.7	73.73	33.93	56.11	1003
42.4096	22181	3.5696	-5257.7	-5139.8	75.33	33.48	55.83	987
42.0756	20589	3.4483	-5147.0	-5028.2	76.92	33.13	56.62	962
41.7428	19812	3.3416	-5036.9	-4917.1	78.45	32.92	56.86	949
41.4061	19097	3.2330	-4925.9	-4805.1	79.95	32.52	56.79	937
41.0695	18234	3.1119	-4815.4	-4693.7	81.41	32.10	56.66	920
40.7274	17393	3.0044	-4704.5	-4581.7	82.83	31.82	56.85	904
40.3746	16273	2.8986	-4592.1	-4468.3	84.24	31.45	57.43	884
40.0301	15907	2.7969	-4481.2	-4356.3	85.60	31.09	56.87	875
39.6779	15107	2.6902	-4369.3	-4243.3	86.93	30.71	56.88	858
39.3213	14293	2.5845	-4256.9	-4129.7	88.24	30.30	56.90	840
38.9607	13615	2.4817	-4145.0	-4016.6	89.52	30.30	57.12	822
38.5918	12850	2.3891	-4031.6	-3902.0	90.78	30.23	57.67	803
38.2163	12143	2.2979	-3917.4	-3786.6	92.02	30.03	58.02	786
37.8376	11529	2.2100	-3803.0	-3670.8	93.24	29.72	58.12	770
37.4501	10883	2.1239	-3687.6	-3554.1	94.44	29.66	58.63	752
37.0580	10258	2.0352	-3571.5	-3436.5	95.62	29.45	58.86	734
36.6557	9626	1.9553	-3454.3	-3317.9	96.79	29.31	59.46	717
36.2452	9012	1.8747	-3335.8	-3197.9	97.95	29.18	60.05	699
35.8282	8487	1.7941	-3216.3	-3076.8	99.09	28.89	60.21	682
35.4016	7921	1.7156	-3095.8	-2954.5	100.22	28.67	60.69	664
34.9634	7369	1.6390	-2973.9	-2830.9	101.33	28.63	61.43	645
34.5097	6795	1.5660	-2849.6	-2704.7	102.45	28.48	62.43	626
34.0445	6328	1.4927	-2723.4	-2577.0	103.56	28.36	62.99	608
33.5636	5817	1.4217	-2595.9	-2446.9	104.66	28.21	64.00	589
33.0671	5345	1.3504	-2465.8	-2314.6	105.77	28.09	64.91	570
32.5496	4861	1.2803	-2332.8	-2179.2	106.87	27.97	66.16	550
32.0093	4379	1.2109	-2196.8	-2040.6	107.98	27.92	67.78	529
31.4438	3941	1.1422	-2057.5	-1898.5	109.09	27.90	69.42	508
30.8459	3497	1.0747	-1914.1	-1752.0	110.22	28.05	71.79	485
30.2092	3073	1.0059	-1764.7	-1599.2	111.37	27.93	74.11	463
29.5343	2664	0.9362	-1610.7	-1441.4	112.54	27.84	76.88	440
28.7990	2262	0.8672	-1448.5	-1274.9	113.74	27.67	80.57	416
27.9916	1868	0.7966	-1277.8	-1099.2	114.99	27.82	85.92	390
27.0801	1479	0.7227	-1049.4	-864.8	116.32	27.92	93.42	361
26.0241	1114	0.6485	-861.1	-669.0	117.75	28.04	104.95	331
24.7137	743	0.5654	-644.3	-442.0	119.38	29.00	127.69	293
22.7963	357	0.4555	-360.7	-141.4	121.51	30.16	188.63	243
19.9891	65	0.3407	-14.2	235.9	124.16	37.26	679.76	176
10.3573	37	0.1498	1179.1	1661.8	134.11	38.38	841.03	147
9.02334	108	0.1249	1416.2	1970.3	136.25	34.21	288.84	155
7.67269	229	0.0993	1699.3	2351.0	138.88	30.31	136.94	165
6.98480	315	0.0869	1869.8	2585.7	140.48	28.45	101.06	172
6.51104	387	0.0783	2002.3	2770.2	141.72	27.30	83.49	176
6.14910	449	0.0728	2114.4	2927.6	142.76	26.48	73.99	182
5.85075	506	0.0682	2214.7	3069.3	143.69	25.84	67.10	186
5.60249	559	0.0641	2305.6	3198.0	144.52	25.32	61.89	190
5.38845	608	0.0607	2389.8	3317.7	145.28	24.89	57.94	193
5.20038	653	0.0579	2468.9	3430.4	145.99	24.54	54.85	196
4.81132	758	0.0521	2650.6	3689.9	147.59	23.88	49.38	203
4.50177	853	0.0477	2816.6	3927.3	149.01	23.43	45.81	209
4.24553	940	0.0442	2972.0	4149.7	150.30	23.11	43.30	215
4.02757	1021	0.0413	3119.9	4361.3	151.49	22.89	41.44	221
3.83847	1099	0.0389	3262.2	4564.8	152.60	22.73	40.01	226
3.67190	1172	0.0368	3400.2	4761.9	153.65	22.62	38.89	230
3.52342	1243	0.0350	3535.0	4954.1	154.65	22.55	37.99	235
3.38977	1311	0.0333	3667.1	5142.1	155.60	22.50	37.25	239
3.26850	1377	0.0319	3797.0	5326.8	156.52	22.48	36.65	243
3.15770	1442	0.0306	3925.3	5508.7	157.39	22.48	36.14	247
3.05588	1504	0.0294	4052.2	5688.4	158.24	22.49	35.72	251
2.96184	1565	0.0283	4177.9	5866.4	159.06	22.51	35.36	254
2.87460	1625	0.0273	4302.7	6042.1	159.85	22.54	35.06	258
2.79335	1684	0.0264	4426.8	6216.8	160.62	22.58	34.81	261
2.71741	1741	0.0256	4550.3	6390.3	161.36	22.63	34.59	265
2.64623	1798	0.0248	4673.3	6562.8	162.09	22.68	34.41	268
2.57930	1854	0.0241	4795.9	6734.5	162.80	22.74	34.26	271
2.51622	1909	0.0234	4918.3	6905.4	163.49	22.81	34.13	274
2.45662	1963	0.0228	5040.5	7075.8	164.16	22.88	34.02	277
2.40020	2017	0.0222	5162.5	7245.7	164.82	22.95	33.94	280
2.34667	2070	0.0216	5284.5	7415.2	165.47	23.03	33.87	283
2.29581	2122	0.0211	5406.5	7584.4	166.10	23.11	33.81	286
2.24738	2174	0.0206	5528.5	7753.3	166.72	23.19	33.77	289
2.20122	2226	0.0201	5650.6	7922.1	167.33	23.27	33.73	291
2.15714	2277	0.0196	5772.8	8090.7	167.92	23.36	33.71	294
2.11500	2327	0.0192	5895.2	8259.2	168.51	23.45	33.70	297
2.07466	2377	0.0188	6017.7	8427.7	169.09	23.54	33.69	299
2.03600	2427	0.0184	6140.4	8596.2	169.65	23.63	33.70	302

E BOUNDARY

TABLE 16. THERMODYNAMIC PROPERTIES OF FLUORINE

5.5 MN/m² ISOBAR

TEMPERATURE (IPTS 1968) K	DENSITY MOL/L	ISOTHERM DERIVATIVE J/MOL	ISOCHORE DERIVATIVE MN/m²-K	INTERNAL ENERGY J/MOL	ENTHALPY J/MOL	ENTROPY J/MOL-K	C_V J / MOL - K	C_p J / MOL - K	VELOCITY OF SOUND M/S
* 54.017	44.9882	28787	4.3322	-6142.2	-6019.9	60.96	36.33	53.73	1059
56	44.6687	27704	4.3287	-6035.0	-5911.9.	62.94	35.99	54.97	1055
58	44.3486	26912	4.2823	-5926.4	-5802.4	64.87	35.62	55.71	1053
60	44.0350	26648	4.1897	-5818.5	-5693.6	66.73	35.32	55.70	1052
62	43.7164	25627	4.0690	-5709.8	-5584.0	68.54	35.23	56.19	1037
64	43.3925	24186	3.9516	-5600.2	-5473.4	70.31	34.97	56.91	1018
66	43.0670	23310	3.8515	-5490.3	-5362.5	72.03	34.57	57.21	1008
68	42.7510	23160	3.7157	-5381.2	-5252.6	73.68	33.96	56.14	1004
70	42.4321	22250	3.5748	-5272.5	-5142.9	75.29	33.49	55.82	988
72	42.0998	20638	3.4538	-5162.2	-5031.5	76.87	33.15	56.63	963
74	41.7680	19826	3.3443	-5052.3	-4920.7	78.40	32.96	56.89	949
76	41.4323	19120	3.2357	-4941.5	-4808.8	79.90	32.57	56.82	937
78	41.0969	18284	3.1165	-4831.3	-4697.5	81.36	32.14	56.67	921
80	40.7561	17446	3.0076	-4720.7	-4585.8	82.78	31.86	56.84	.905
82	40.4053	16312	2.9017	-4608.8	-4472.7	84.19	31.50	57.43	885
84	40.0614	15969	2.8008	-4498.1	-4360.8	85.54	31.14	56.85	876
86	39.7109	15183	2.6956	-4386.5	-4248.0	86.88	30.75	56.85	859
88	39.3562	14379.	2.5909	-4274.5	-4134.7	88.18	30.34	56.86	842
90	38.9973	13720	2.4884	-4162.9	-4021.9	89.46	30.33	57.04	824
92	38.6306	12947	2.3947	-4050.0	-3907.6	90.72	30.29	57.60	805
94	38.2573	12233	2.3030	-3936.3	-3792.5	91.96	30.09	57.94	787
96	37.8808	11631	2.2148	-3822.3	-3677.1	93.17	29.77	57.99	772
98	37.4958	10984	2.1301	-3707.4	-3560.7	94.37	29.72	58.52	754
100	37.1065	10367	2.0419	-3591.8	-3443.5	95.55	29.52	58.73	737
102	36.7074	9732	1.9614	-3475.2	-3325.4	96.71	29.37	59.29	719
104	36.3004	9116	1.8819	-3357.4	-3205.9	97.87	29.25	59.91	701
106	35.8867	8602	1.8015	-3238.5	-3085.2	99.01	28.95	60.00	685
108	35.4643	8044	1.7239	-3118.6	-2963.5	100.13	28.72	60.44	668
110	35.0306	7496	1.6482	-2997.5	-2840.5	101.24	28.67	61.15	649
112	34.5826	6922	1.5755	-2874.2	-2715.2	102.36	28.53	62.12	630
114	34.1227	6454	1.5032	-2749.4	-2588.2	103.46	28.40	62.68	612
116	33.6486	5947	1.4326	-2622.5	-2459.0	104.56	28.25	63.61	594
118	33.1595	5480	1.3624	-2493.5	-2327.6	105.65	28.12	64.47	575
120	32.6510	5001	1.2930	-2361.9	-2193.5	106.75	28.00	65.63	555
122	32.1216	4522	1.2247	-2227.5	-2056.3	107.85	27.93	67.15	535
124	31.5684	4087	1.1566	-2090.0	-1915.7	108.95	27.91	68.64	514
126	30.9859	3646	1.0899	-1948.7	-1771.2	110.06	28.06	70.81	492
128	30.3680	3225	1.0223	-1801.8	-1620.7	111.20	27.96	72.93	471
130	29.7165	2823	0.9548	-1650.9	-1465.8	112.34	27.83	75.36	449
132	29.0123	2428	0.8875	-1492.7	-1303.1	113.52	27.64	78.51	426
134	28.2475	2042	0.8192	-1327.2	-1132.5	114.73	27.73	82.93	401
136	27.3989	1661	0.7498	-1090.2	-889.4	116.01	27.81	89.13	374
138	26.4371	1310	0.6793	-912.6	-704.5	117.36	27.78	97.33	348
140	25.3035	958	0.6034	-715.9	-498.6	118.84	28.30	111.41	315
142	23.8512	604	0.5162	-483.2	-252.6	120.58	29.01	139.14	276
144	21.6567	267	0.4098	-171.2	82.8	122.93	30.95	223.92	226
146	11.8523	47	0.1723	1058.1	1522.1	132.83	37.91	692.23	151
148	9.04992	179	0.1215	1542.8	2150.5	137.11	31.35	180.35	165
150	8.04361	272	0.1028	1759.0	2442.7	139.07	29.07	119.29	171
152	7.41155	347	0.0925	1914.3	2656.4	140.49	27.74	95.99	178
154	6.94438	414	0.0851	2041.7	2833.7	141.65	26.81	82.73	183
156	6.57626	473	0.0782	2152.0	2988.4	142.65	26.13	72.73	186
158	6.27296	528	0.0737	2250.9	3127.7	143.53	25.58	66.84	191
160	6.01379	579	0.0698	2341.8	3256.4	144.34	25.14	62.39	195
165	5.49979	695	0.0614	2544.5	3544.5	146.12	24.29	53.88	201
170	5.10815	797	0.0556	2724.1	3800.9	147.65	23.74	49.01	208
175	4.79212	890	0.0511	2889.5	4037.2	149.02	23.36	45.73	214
180	4.52814	977	0.0475	3045.0	4259.7	150.27	23.09	43.37	220
185	4.30220	1058	0.0445	3193.5	4471.9	151.44	22.90	41.60	225
190	4.10529	1136	0.0419	3336.6	4676.3	152.53	22.77	40.22	230
195	3.93124	1209	0.0397	3475.5	4874.6	153.56	22.68	39.13	234
200	3.77567	1280	0.0378	3611.3	5068.0	154.54	22.62	38.25	239
205	3.63532	1349	0.0360	3744.4	5257.4	155.47	22.59	37.53	243
210	3.50774	1415	0.0345	3875.5	5443.5	156.37	22.57	36.93	247
215	3.39099	1480	0.0331	4004.9	5626.8	157.23	22.57	36.43	251
220	3.28357	1543	0.0318	4132.8	5807.9	158.06	22.59	36.00	254
225	3.18423	1605	0.0307	4259.7	5987.0	158.87	22.62	35.65	258
230	3.09198	1665	0.0296	4385.6	6164.4	159.65	22.65	35.34	261
235	3.00598	1724	0.0287	4510.8	6340.5	160.41	22.70	35.09	265
240	2.92555	1782	0.0278	4635.3	6515.3	161.14	22.75	34.87	268
245	2.85008	1839	0.0269	4759.4	6689.2	161.86	22.80	34.68	271
250	2.77907	1896	0.0261	4883.1	6862.2	162.56	22.87	34.53	274
255	2.71210	1951	0.0254	5006.5	7034.5	163.24	22.93	34.39	278
260	2.64879	2006	0.0247	5129.7	7206.2	163.91	23.00	34.28	280
265	2.58882	2060	0.0241	5252.8	7377.3	164.56	23.08	34.19	283
270	2.53190	2113	0.0235	5375.8	7548.1	165.20	23.16	34.12	286
275	2.47778	2166	0.0229	5498.8	7718.5	165.82	23.24	34.06	289
280	2.42623	2219	0.0223	5621.8	7888.7	166.44	23.32	34.01	292
285	2.37706	2271	0.0218	5744.9	8058.7	167.04	23.40	33.97	295
290	2.33010	2322	0.0213	5868.1	8228.5	167.63	23.49	33.95	297
295	2.28518	2373	0.0209	5991.4	8398.2	168.21	23.58	33.93	300
300	2.24216	2424	0.0204	6114.8	8567.8	168.78	23.66	33.92	302

* TWO-PHASE BOUNDARY

TABLE 16. THERMODYNAMIC PROPERTIES OF FLUORINE

/m² ISOBAR

DENSITY MOL/L	ISOTHERM DERIVATIVE J/MOL	ISOCHORE DERIVATIVE MN/m²-K	INTERNAL ENERGY J/MOL	ENTHALPY J/MOL	ENTROPY J/MOL-K	C_v J / MOL - K	C_p J / MOL - K	VELOCITY OF SOUND M/S
44.9977	28915	4.3337	-6152.6	-6019.2	60.97	36.32	53.66	1060
44.6867	27829	4.3302	-6048.3	-5914.0	62.90	35.98	54.87	1057
44.3671	26998	4.2889	-5939.9	-5804.7	64.83	35.61	55.69	1054
44.0537	26730	4.1963	-5832.2	-5696.0	66.69	35.31	55.68	1053
43.7359	25665	4.0735	-5723.7	-5586.5	68.50	35.25	56.21	1038
43.4132	24183	3.9509	-5614.4	-5476.2	70.27	35.01	56.93	1017
43.0885	23320	3.8502	-5504.7	-5365.4	71.99	34.62	57.21	1007
42.7726	23206	3.7199	-5395.8	-5255.5	73.64	33.99	56.15	1004
42.4545	22321	3.5801	-5287.2	-5145.9	75.24	33.51	55.81	989
42.1240	20692	3.4594	-5177.3	-5034.9	76.82	33.17	56.64	964
41.7932	19849	3.3474	-5067.7	-4924.2	78.35	33.00	56.91	949
41.4584	19149	3.2382	-4957.2	-4812.4	79.85	32.62	56.84	937
41.1242	18338	3.1207	-4847.2	-4701.3	81.31	32.18	56.67	922
40.7847	17504	3.0109	-4736.9	-4589.8	82.73	31.91	56.82	906
40.4359	16359	2.9047	-4625.4	-4477.0	84.13	31.56	57.42	885
40.0927	16033	2.8046	-4514.9	-4365.2	85.49	31.19	56.82	877
39.7437	15262	2.7008	-4403.7	-4252.7	86.82	30.80	56.82	861
39.3908	14466	2.5970	-4292.0	-4139.7	88.13	30.38	56.82	844
39.0336	13823	2.4961	-4180.8	-4027.1	89.40	30.36	56.98	826
38.6691	13046	2.4008	-4068.4	-3913.2	90.65	30.34	57.52	807
38.2981	12326	2.3083	-3955.1	-3798.4	91.89	30.15	57.85	789
37.9236	11734	2.2197	-3841.4	-3683.2	93.11	29.83	57.86	774
37.5412	11086	2.1360	-3727.1	-3567.3	94.30	29.78	58.40	756
37.1545	10476	2.0488	-3612.0	-3450.5	95.48	29.58	58.60	739
36.7585	9839	1.9675	-3496.0	-3332.8	96.64	29.42	59.12	721
36.3549	9222	1.8892	-3378.8	-3213.7	97.79	29.30	59.76	704
35.9444	8718	1.8087	-3260.4	-3093.5	98.92	29.00	59.79	688
35.5260	8167	1.7323	-3141.3	-2972.4	100.05	28.76	60.20	671
35.0968	7623	1.6578	-3020.9	-2850.0	101.15	28.70	60.90	652
34.6542	7049	1.5850	-2898.5	-2725.4	102.26	28.58	61.82	633
34.1995	6578	1.5134	-2774.5	-2599.1	103.36	28.43	62.37	616
33.7318	6077	1.4433	-2648.7	-2470.8	104.45	28.29	63.24	598
33.2496	5613	1.3741	-2520.8	-2340.4	105.54	28.15	64.05	580
32.7497	5137	1.3053	-2390.5	-2207.3	106.63	28.02	65.13	>61
32.2305	4664	1.2379	-2257.6	-2071.5	107.72	27.95	66.54	541
31.6886	4231	1.1709	-2121.7	-1932.4	108.81	27.93	67.94	520
31.1204	3793	1.1049	-1982.4	-1789.6	109.91	28.07	69.94	499
30.5195	3374	1.0384	-1837.9	-1641.4	111.03	27.97	71.88	478
29.8889	2979	0.9716	-1689.7	-1489.0	112.15	27.83	73.95	456
29.2116	2589	0.9067	-1535.0	-1329.6	113.31	27.61	76.73	435
28.4828	2209	0.8398	-1374.0	-1163.4	114.49	27.65	80.38	411
27.6849	1837	0.7743	-1127.1	-910.4	115.72	27.72	85.64	386
26.7941	1493	0.7035	-957.6	-733.7	117.01	27.61	91.33	361
25.7780	1153	0.6351	-774.6	-541.8	118.39	27.92	101.63	332
24.5584	817	0.5603	-568.2	-323.9	119.93	28.36	116.84	300
22.9795	506	0.4742	-322.8	-61.7	121.77	29.06	150.33	262
20.5202	215	0.3689	19.2	311.6	124.34	31.67	251.35	212
14.2507	55	0.2125	814.0	1235.1	135.61	36.83	630.39	158
10.4630	154	0.1439	1395.2	1968.7	135.75	31.87	216.20	166
9.12224	244	0.1208	1653.2	2310.9	137.81	29.43	138.71	174
8.31350	321	0.1073	1831.7	2553.4	139.40	28.00	107.83	180
7.74012	389	0.0963	1973.3	2748.5	140.66	27.02	89.06	184
7.29777	450	0.0893	2093.6	2915.8	141.72	26.30	78.84	188
6.93624	507	0.0837	2200.4	3065.5	142.67	25.74	71.70	193
6.25485	632	0.0726	2429.6	3388.9	144.66	24.74	59.87	201
5.75736	744	0.0645	2626.4	3668.5	146.33	24.07	52.73	207
5.36932	843	0.0587	2803.3	3920.8	147.79	23.61	48.46	213
5.05161	934	0.0542	2967.5	4155.3	149.11	23.29	45.49	219
4.78362	1019	0.0505	3122.7	4377.0	150.33	23.07	43.31	224
4.55267	1100	0.0474	3271.3	4589.3	151.46	22.92	41.65	229
4.35034	1177	0.0448	3414.9	4794.1	152.52	22.81	40.34	234
4.17079	1251	0.0424	3554.5	4993.1	153.53	22.74	39.30	239
4.00978	1322	0.0404	3691.1	5187.4	154.49	22.69	38.44	243
3.86415	1390	0.0386	3825.1	5377.8	155.41	22.67	37.74	247
3.73147	1457	0.0370	3957.0	5565.0	156.29	22.66	37.15	251
3.60983	1522	0.0355	4087.4	5749.5	157.14	22.67	36.66	254
3.49772	1585	0.0342	4216.3	5931.7	157.96	22.69	36.24	258
3.39390	1647	0.0330	4344.1	6112.0	158.75	22.72	35.89	262
3.29737	1708	0.0318	4471.0	6290.7	159.52	22.76	35.59	265
3.20728	1767	0.0308	4597.2	6467.9	160.26	22.81	35.33	268
3.12292	1826	0.0298	4722.7	6644.0	150.99	22.86	35.11	272
3.04370	1883	0.0289	4847.8	6819.1	161.70	22.92	34.92	275
2.96911	1940	0.0281	4972.5	6993.3	162.39	22.99	34.76	278
2.89870	1996	0.0273	5096.9	7166.8	163.06	23.06	34.63	281
2.83208	2051	0.0266	5221.0	7339.6	163.72	23.13	34.52	284
2.76893	2105	0.0259	5345.1	7512.0	164.36	23.20	34.43	287
2.70896	2159	0.0253	5469.0	7683.9	165.00	23.28	34.35	290
2.65190	2213	0.0246	5593.0	7855.5	165.61	23.36	34.29	292
2.59753	2265	0.0241	5716.9	8026.8	166.22	23.44	34.24	295
2.54563	2318	0.0235	5840.9	8197.9	166.81	23.53	34.20	298
2.49604	2369	0.0230	5965.0	8368.8	167.40	23.61	34.17	300
2.44858	2421	0.0225	6089.2	8539.6	167.97	23.70	34.15	303

E BOUNDARY

TABLE 16. THERMODYNAMIC PROPERTIES OF FLUORINE

6.5 MN/m² ISOBAR

TEMPERATURE (IPTS 1968) K	DENSITY MOL/L	ISOTHERM DERIVATIVE J/MOL	ISOCHORE DERIVATIVE MN/m²-K	INTERNAL ENERGY J/MOL	ENTHALPY J/MOL	ENTROPY J/MOL-K	C_v J / MOL - K	C_p J / MOL - K	VELOCITY OF SOUND M/S
* 54.113	45.0072	29041	4.3352	-6163.0	-6018.6	60.98	36.30	53.59	1062
56	44.7047	27955	4.3317	-6061.5	-5916.1	62.86	35.97	54.78	1058
58	44.3856	27063	4.2955	-5953.4	-5807.0	64.79	35.61	55.66	1056
60	44.0724	26811	4.2029	-5845.9	-5698.4	66.65	35.31	55.66	1055
62	43.7554	25709	4.0786	-5737.6	-5589.1	68.46	35.27	56.22	1039
64	43.4339	24190	3.9504	-5628.5	-5478.9	70.22	35.05	56.93	1017
66	43.1099	23337	3.8490	-5519.1	-5368.3	71.94	34.66	57.21	1007
68	42.7941	23254	3.7237	-5410.2	-5258.4	73.60	34.02	56.16	1005
70	42.4769	22394	3.5654	-5302.0	-5149.0	75.20	33.53	55.80	990
72	42.1481	20753	3.4650	-5192.4	-5038.2	76.77	33.19	56.64	965
74	41.8184	19881	3.3508	-5083.1	-4927.7	78.30	33.03	56.93	950
76	41.4845	19185	3.2405	-4972.8	-4816.1	79.80	32.67	56.85	937
78	41.1514	18397	3.1248	-4863.1	-4705.1	81.26	32.22	56.67	923
80	40.8132	17566	3.0144	-4753.1	-4593.9	82.68	31.95	56.80	906
82	40.4664	16414	2.9076	-4642.0	-4481.3	84.08	31.61	57.40	886
84	40.1238	16100	2.8082	-4531.7	-4369.7	85.43	31.24	56.79	878
86	39.7764	15342	2.7058	-4420.8	-4257.4	86.76	30.84	56.78	862
88	39.4253	14556	2.6028	-4309.5	-4144.6	88.07	30.42	56.77	846
90	39.0697	13926	2.5036	-4198.6	-4032.3	89.34	30.38	56.92	829
92	38.7073	13147	2.4072	-4086.6	-3918.7	90.59	30.39	57.46	809
94	38.3385	12422	2.3139	-3973.8	-3804.2	91.83	30.20	57.77	791
96	37.9660	11838	2.2248	-3860.5	-3689.3	93.04	29.88	57.73	776
98	37.5860	11190	2.1419	-3746.7	-3573.7	94.23	29.83	58.27	758
100	37.2020	10586	2.0558	-3632.0	-3457.3	95.41	29.64	58.48	741
102	36.8090	9948	1.9737	-3516.6	-3340.0	96.57	29.47	58.95	724
104	36.4089	9330	1.8966	-3400.0	-3221.5	97.71	29.36	59.61	706
106	36.0014	8833	1.8160	-3282.3	-3101.7	98.84	29.05	59.99	690
108	35.5867	8289	1.7408	-3163.7	-2981.1	99.96	28.81	59.99	674
110	35.1618	7749	1.6671	-3044.2	-2859.3	101.07	28.74	60.65	656
112	34.7245	7178	1.5945	-2922.6	-2735.4	102.17	28.62	61.52	637
114	34.2748	6701	1.5234	-2799.5	-2609.8	103.26	28.47	62.08	620
116	33.8132	6205	1.4538	-2674.6	-2482.3	104.35	28.32	62.88	602
118	33.3377	5743	1.3853	-2547.8	-2352.8	105.43	28.18	63.66	584
120	32.8457	5271	1.3173	-2418.7	-2220.8	106.51	28.05	64.66	566
122	32.3361	4804	1.2507	-2287.2	-2086.2	107.59	27.97	65.96	546
124	31.8049	4372	1.1847	-2152.9	-1948.5	108.67	27.94	67.29	526
126	31.2497	3938	1.1194	-2015.4	-1807.4	109.76	28.07	69.13	505
128	30.6646	3521	1.0543	-1873.1	-1661.1	110.87	27.97	70.94	485
130	30.0526	3131	0.9877	-1727.3	-1511.0	111.98	27.84	72.70	464
132	29.3991	2745	0.9248	-1575.7	-1354.6	113.11	27.59	75.17	444
134	28.7011	2372	0.8604	-1418.6	-1192.1	114.26	27.59	78.36	421
136	27.9453	2006	0.7965	-1160.9	-928.3	115.45	27.64	82.71	398
138	27.1107	1667	0.7292	-998.0	-756.2	116.70	27.48	87.37	374
140	26.1804	1335	0.6633	-824.9	-576.6	118.00	27.66	94.96	347
142	25.1066	1012	0.5963	-635.4	-376.5	119.42	27.99	107.15	319
144	23.8061	712	0.5196	-420.6	-147.6	121.02	28.30	124.64	287
146	22.1136	433	0.4378	-163.3	130.6	122.94	29.46	161.54	250
148	19.5430	200	0.3409	193.3	525.9	125.63	31.48	256.41	207
150	15.0599	94	0.2296	784.5	1216.1	130.26	33.98	403.14	172
152	11.7077	155	0.1658	1284.0	1839.2	134.59	31.54	228.35	172
154	10.1336	235	0.1382	1564.7	2206.1	136.79	29.43	151.22	178
156	9.16237	309	0.1199	1759.2	2467.0	138.47	28.07	114.20	182
158	8.51353	376	0.1088	1912.1	2675.6	139.80	27.12	95.80	187
160	7.99852	437	0.1004	2041.4	2854.1	140.92	26.40	84.09	191
165	7.08628	572	0.0850	2305.2	3222.5	143.19	25.19	66.63	200
170	6.45711	690	0.0747	2522.3	3528.9	145.02	24.41	57.44	207
175	5.97947	797	0.0671	2713.3	3800.4	146.60	23.88	51.53	213
180	5.59919	893	0.0615	2887.3	4048.2	147.99	23.50	47.82	219
185	5.28338	982	0.0570	3050.0	4280.3	149.27	23.25	45.16	224
190	5.01438	1066	0.0533	3204.6	4500.9	150.44	23.07	43.17	229
195	4.78087	1146	0.0501	3353.1	4712.7	151.54	22.94	41.62	234
200	4.57518	1222	0.0474	3496.9	4917.6	152.58	22.85	40.39	238
205	4.39185	1296	0.0450	3637.0	5117.0	153.57	22.80	39.40	243
210	4.22687	1367	0.0429	3774.1	5311.8	154.51	22.77	38.58	247
215	4.07722	1435	0.0410	3908.8	5503.0	155.40	22.75	37.90	251
220	3.94054	1502	0.0393	4041.5	5691.0	156.27	22.75	37.33	255
225	3.81497	1567	0.0378	4172.6	5876.4	157.10	22.77	36.85	258
230	3.69902	1630	0.0364	4302.4	6059.6	157.91	22.80	36.44	262
235	3.59147	1692	0.0351	4431.1	6240.9	158.69	22.83	36.09	265
240	3.49133	1753	0.0339	4558.9	6420.6	159.44	22.87	35.79	269
245	3.39775	1813	0.0328	4685.9	6598.9	160.18	22.92	35.54	272
250	3.31003	1872	0.0318	4812.4	6776.1	160.90	22.98	35.32	275
255	3.22755	1929	0.0309	4938.3	6952.2	161.59	23.04	35.14	278
260	3.14982	1986	0.0300	5063.9	7127.5	162.27	23.10	34.98	281
265	3.07638	2043	0.0292	5189.2	7302.1	162.94	23.17	34.85	284
270	3.00684	2098	0.0284	5314.3	7476.0	163.59	23.25	34.73	287
275	2.94087	2153	0.0277	5439.2	7649.4	164.23	23.32	34.64	290
280	2.87817	2207	0.0270	5564.0	7822.4	164.85	23.40	34.56	293
285	2.81848	2261	0.0263	5688.8	7995.1	165.46	23.48	34.50	296
290	2.76156	2314	0.0257	5813.7	8167.4	166.06	23.56	34.44	298
295	2.70721	2366	0.0251	5938.5	8339.5	166.65	23.65	34.41	301
300	2.65523	2418	0.0246	6063.5	8511.5	167.23	23.73	34.38	304

* TWO-PHASE BOUNDARY

TABLE 16. THERMODYNAMIC PROPERTIES OF FLUORINE

7.0 MN/m² ISOBAR

PERATURE TS 1968) K	DENSITY MOL/L	ISOTHERM DERIVATIVE J/MOL	ISOCHORE DERIVATIVE MN/m²-K	INTERNAL ENERGY J/MOL	ENTHALPY J/MOL	ENTROPY J/MOL-K	C_v J / MOL - K	C_p J / MOL - K	VELOCITY OF SOUND M/S
54.162	45.0167	29165	4.3367	-6173.4	-6017.9	60.99	36.29	53.53	1064
56	44.7225	28079	4.3331	-6074.7	-5918.2	62.82	35.97	54.69	1060
58	44.4041	27169	4.3020	-5966.9	-5809.2	64.75	35.60	55.64	1057
60	44.0910	26892	4.2094	-5859.5	-5700.7	66.61	35.30	55.64	1056
62	43.7748	25758	4.0842	-5751.5	-5591.6	58.42	35.29	56.24	1039
64	43.4545	24208	3.9500	-5642.6	-5481.6	70.18	35.08	56.93	1017
66	43.1313	23363	3.8480	-5533.4	-5371.1	71.90	34.71	57.19	1007
68	42.8156	23305	3.7271	-5424.7	-5261.2	73.55	34.05	56.16	1006
70	42.4992	22469	3.5907	-5316.7	-5152.0	75.15	33.55	55.79	992
72	42.1722	20819	3.4706	-5207.4	-5041.5	76.73	33.22	56.64	967
74	41.8435	19923	3.3546	-5098.4	-4931.1	78.25	33.07	56.94	950
76	41.5105	19226	3.2425	-4988.3	-4819.7	79.75	32.72	56.84	938
78	41.1786	18461	3.1286	-4878.9	-4708.9	81.21	32.26	56.66	924
80	40.8416	17632	3.0179	-4769.3	-4597.9	82.63	32.00	56.77	907
82	40.4968	16477	2.9104	-4658.5	-4485.6	84.02	31.66	57.37	886
84	40.1548	16170	2.8117	-4548.4	-4374.1	85.38	31.28	56.75	879
86	39.8089	15425	2.7106	-4437.8	-4262.0	86.71	30.89	56.74	864
88	39.4595	14649	2.6084	-4326.9	-4149.5	88.01	30.46	56.71	847
90	39.1054	14029	2.5107	-4216.4	-4037.4	89.28	30.41	56.86	831
92	38.7451	13249	2.4140	-4104.8	-3924.1	90.53	30.44	57.39	811
94	38.3786	12520	2.3196	-3992.4	-3810.0	91.76	30.25	57.68	793
96	38.0080	11942	2.2302	-3879.5	-3695.4	92.97	29.93	57.61	778
98	37.6305	11294	2.1479	-3766.1	-3580.1	94.16	29.87	58.14	761
100	37.2490	10696	2.0630	-3652.0	-3464.1	95.34	29.69	58.37	744
102	36.8590	10057	1.9800	-3537.1	-3347.2	96.49	29.52	58.78	726
104	36.4621	9440	1.9042	-3421.1	-3229.2	97.64	29.41	59.45	709
106	36.0576	8948	1.8242	-3303.9	-3109.8	98.77	29.10	59.43	693
108	35.6466	8410	1.7493	-3186.0	-2989.7	99.88	28.85	59.77	677
110	35.2258	7874	1.6762	-3067.2	-2868.4	100.98	28.77	60.41	660
112	34.7935	7307	1.6041	-2946.5	-2745.3	102.08	28.66	61.24	641
114	34.3487	6823	1.5331	-2824.1	-2620.3	103.17	28.50	61.79	624
116	33.8930	6332	1.4642	-2700.2	-2493.7	104.25	28.35	62.54	606
118	33.4238	5871	1.3963	-2574.4	-2365.0	105.32	28.21	63.28	589
120	32.9394	5404	1.3290	-2446.5	-2234.0	106.40	28.07	64.22	570
122	32.4387	4943	1.2630	-2316.3	-2100.6	107.47	27.98	65.40	551
124	31.9174	4511	1.1980	-2183.4	-1964.1	108.54	27.95	66.67	532
126	31.3744	4081	1.1333	-2047.7	-1824.6	109.62	28.08	68.37	511
128	30.8037	3666	1.0694	-1907.4	-1680.2	110.71	27.97	70.06	492
130	30.2086	3280	1.0044	-1763.8	-1532.1	111.81	27.86	71.68	471
132	29.5764	2898	0.9421	-1615.0	-1378.4	112.92	27.56	73.77	452
134	28.9052	2530	0.8792	-1461.3	-1219.1	114.05	27.55	76.56	430
136	28.1847	2171	0.8173	-1192.2	-943.9	115.21	27.57	80.25	408
138	27.3965	1834	0.7532	-1034.7	-779.2	116.41	27.38	84.25	385
140	26.5323	1509	0.6880	-869.4	-605.5	117.66	27.45	89.85	360
142	25.5604	1195	0.6272	-691.8	-417.9	118.99	27.73	99.29	336
144	24.4278	902	0.5566	-496.0	-209.5	120.45	27.94	110.87	307
146	23.0625	630	0.4856	-275.0	28.5	122.09	28.51	131.24	276
148	21.2920	391	0.4069	-8.4	320.4	124.07	29.63	167.69	241
150	18.7769	210	0.3226	343.8	716.6	126.73	30.85	241.94	208
152	15.3569	139	0.2358	811.9	1267.7	130.38	31.54	288.55	183
154	12.6285	176	0.1825	1222.8	1777.1	133.71	30.62	213.05	180
156	11.0175	242	0.1517	1499.4	2134.8	136.02	29.10	151.07	182
158	9.98193	310	0.1335	1700.7	2402.0	137.72	27.95	119.21	186
160	9.23686	374	0.1208	1861.5	2619.3	139.09	27.08	100.34	191
165	8.00665	516	0.0993	2170.3	3044.5	141.71	25.65	74.81	199
170	7.20996	640	0.0859	2411.9	3382.7	143.73	24.76	62.51	206
175	6.62813	750	0.0766	2618.6	3674.7	145.43	24.15	55.31	213
180	6.17322	851	0.0697	2804.1	3938.0	146.91	23.72	50.64	219
185	5.80194	947	0.0639	2975.3	4181.8	148.25	23.42	47.14	224
190	5.49060	1034	0.0595	3136.4	4411.3	149.47	23.22	44.77	229
195	5.22287	1117	0.0557	3290.2	4630.4	150.61	23.07	42.96	234
200	4.98879	1196	0.0525	3438.4	4841.5	151.68	22.97	41.53	239
205	4.78144	1271	0.0498	3582.2	5046.2	152.69	22.90	40.38	243
210	4.59580	1344	0.0473	3722.5	5245.6	153.65	22.86	39.44	247
215	4.42813	1415	0.0452	3860.0	5440.8	154.57	22.84	38.66	251
220	4.27556	1483	0.0432	3995.3	5632.5	155.45	22.84	38.01	255
225	4.13585	1550	0.0415	4128.6	5821.2	156.30	22.85	37.47	259
230	4.00721	1615	0.0399	4260.4	6007.3	157.12	22.87	37.00	262
235	3.88818	1678	0.0385	4391.0	6191.3	157.91	22.90	36.60	266
240	3.77760	1740	0.0371	4520.4	6373.4	158.67	22.93	36.27	269
245	3.67446	1801	0.0359	4649.0	6554.0	159.42	22.98	35.98	272
250	3.57794	1861	0.0348	4776.8	6733.3	160.14	23.03	35.73	276
255	3.48735	1920	0.0337	4904.1	6911.3	160.85	23.09	35.51	279
260	3.40208	1978	0.0327	5030.9	7088.4	161.54	23.15	35.33	282
265	3.32163	2035	0.0318	5157.3	7264.7	162.21	23.22	35.17	285
270	3.24554	2091	0.0310	5283.4	7440.2	162.86	23.29	35.04	288
275	3.17344	2147	0.0302	5409.3	7615.1	163.51	23.36	34.93	291
280	3.10498	2202	0.0294	5535.1	7789.5	164.13	23.44	34.83	293
285	3.03986	2257	0.0287	5660.7	7963.5	164.75	23.52	34.76	296
290	2.97782	2311	0.0280	5786.4	8137.1	165.35	23.60	34.69	299
295	2.91862	2364	0.0274	5912.0	8310.4	165.95	23.68	34.64	302
300	2.86205	2417	0.0267	6037.7	8483.5	166.53	23.76	34.60	304

TABLE 16. THERMODYNAMIC PROPERTIES OF FLUORINE

7.5 MN/m² ISOBAR

TEMPERATURE (IPTS 1968) K	DENSITY MOL/L	ISOTHERM DERIVATIVE J/MOL	ISOCHORE DERIVATIVE MN/m²-K	INTERNAL ENERGY J/MOL	ENTHALPY J/MOL	ENTROPY J/MOL-K	C_v J / MOL - K	C_p J / MOL - K	VELOCITY OF SOUND M/S
* 54.210	45.0261	29286	4.3380	-6183.7	-6017.2	61.00	36.28	53.46	1066
56	44.7403	28203	4.3344	-6087.9	-5920.3	62.78	35.96	54.60	1062
58	44.4224	27254	4.3084	-5980.3	-5811.5	64.71	35.60	55.62	1059
60	44.1095	26972	4.2158	-5873.1	-5703.1	66.57	35.30	55.62	1058
62	43.7942	25812	4.0905	-5765.4	-5594.1	68.37	35.30	56.26	1040
64	43.4752	24235	3.9497	-5656.8	-5484.2	70.14	35.12	56.92	1017
66	43.1527	23396	3.8471	-5547.8	-5374.0	71.85	34.75	57.17	1006
68	42.8370	23357	3.7299	-5439.2	-5264.1	73.51	34.09	56.16	1006
70	42.5214	22545	3.5971	-5331.4	-5155.0	75.11	33.57	55.79	993
72	42.1962	20891	3.4762	-5222.5	-5044.7	76.68	33.24	56.63	968
74	41.8686	19973	3.3588	-5113.7	-4934.6	78.21	33.10	56.94	951
76	41.5365	19274	3.2446	-5003.9	-4823.3	79.71	32.77	56.84	938
78	41.2056	18528	3.1322	-4894.7	-4712.7	81.16	32.31	56.63	925
80	40.8699	17703	3.0216	-4785.3	-4601.8	82.58	32.04	56.74	908
82	40.5271	16548	2.9139	-4675.0	-4489.9	83.97	31.72	57.33	887
84	40.1857	16242	2.8151	-4565.1	-4378.5	85.33	31.33	56.71	880
86	39.8412	15509	2.7153	-4454.9	-4266.6	86.65	30.93	56.69	865
88	39.4935	14744	2.6136	-4344.3	-4154.4	87.95	30.50	56.64	849
90	39.1409	14132	2.5176	-4234.1	-4042.5	89.22	30.44	56.79	833
92	38.7827	13353	2.4210	-4122.9	-3929.5	90.47	30.48	57.33	813
94	38.4183	12621	2.3256	-4010.9	-3815.7	91.70	30.30	57.59	795
96	38.0497	12048	2.2366	-3898.5	-3701.3	92.91	29.97	57.51	780
98	37.6746	11400	2.1540	-3785.5	-3586.5	94.10	29.92	58.02	763
100	37.2955	10807	2.0702	-3671.8	-3470.7	95.27	29.74	58.25	746
102	36.9085	10168	1.9864	-3557.5	-3354.3	96.42	29.56	58.61	728
104	36.5148	9551	1.9117	-3442.1	-3236.7	97.56	29.45	59.30	711
106	36.1132	9063	1.8325	-3325.4	-3117.7	98.69	29.15	59.27	696
108	35.7057	8531	1.7576	-3208.2	-2998.1	99.80	28.88	59.56	680
110	35.2888	7998	1.6852	-3090.0	-2877.4	100.90	28.80	60.17	663
112	34.8613	7437	1.6137	-2970.1	-2754.9	101.99	28.69	60.96	645
114	34.4214	6944	1.5426	-2848.5	-2630.7	103.07	28.54	61.51	628
116	33.9712	6458	1.4744	-2725.5	-2504.7	104.15	28.38	62.21	610
118	33.5080	5998	1.4070	-2600.7	-2376.8	105.22	28.23	62.92	593
120	33.0308	5535	1.3406	-2473.9	-2246.8	106.28	28.09	63.80	575
122	32.5385	5080	1.2752	-2345.0	-2114.5	107.35	28.00	64.89	557
124	32.0266	4649	1.2110	-2213.5	-1979.3	108.41	27.96	66.10	538
126	31.4949	4222	1.1468	-2079.4	-1841.3	109.48	28.09	67.65	517
128	30.9376	3808	1.0840	-1940.0	-1698.5	110.56	27.98	69.24	498
130	30.3578	3426	1.0206	-1799.4	-1552.3	111.64	27.86	70.76	478
132	29.7446	3048	0.9580	-1653.1	-1400.9	112.74	27.54	72.47	459
134	29.0971	2684	0.8973	-1502.3	-1244.6	113.85	27.52	75.00	439
136	28.4069	2331	0.8368	-1221.4	-957.4	114.98	27.50	78.13	417
138	27.6577	1996	0.7751	-1068.5	-797.4	116.15	27.31	81.62	396
140	26.8465	1675	0.7125	-909.4	-630.0	117.35	27.30	86.16	373
142	25.9508	1369	0.6547	-740.8	-451.8	118.62	27.52	93.54	350
144	24.9334	1080	0.5886	-558.3	-257.5	119.97	27.70	102.01	323
146	23.7582	813	0.5237	-358.6	-42.9	121.45	27.97	115.20	297
148	22.3394	573	0.4535	-131.7	204.0	123.13	28.76	135.20	266
150	20.5476	372	0.3835	137.8	502.8	125.14	29.53	170.02	237
152	18.1998	233	0.3055	472.2	884.3	127.66	30.24	213.70	208
154	15.5117	185	0.2391	854.7	1338.3	130.63	30.42	228.52	191
156	13.2735	209	0.1938	1199.0	1764.0	133.38	29.65	188.48	187
158	11.7458	263	0.1649	1458.3	2096.8	135.50	28.60	147.03	189
160	10.6819	322	0.1457	1659.0	2361.1	137.16	27.69	120.06	192
165	9.02682	466	0.1159	2024.2	2855.0	140.21	26.10	84.43	199
170	8.02101	594	0.0985	2295.0	3230.0	142.45	25.10	68.25	206
175	7.31360	709	0.0868	2519.9	3545.3	144.28	24.43	59.18	213
180	6.77391	814	0.0783	2717.9	3825.1	145.85	23.95	53.46	219
185	6.34093	911	0.0716	2898.4	4091.8	147.26	23.61	49.51	224
190	5.98137	1004	0.0661	3066.7	4320.6	148.53	23.37	46.46	229
195	5.67627	1089	0.0617	3226.1	4547.4	149.71	23.20	44.35	234
200	5.41151	1170	0.0580	3379.0	4764.9	150.81	23.09	42.71	239
205	5.17841	1248	0.0548	3526.7	4975.0	151.85	23.01	41.39	243
210	4.97078	1323	0.0520	3670.4	5179.3	152.84	22.96	40.33	247
215	4.78405	1395	0.0495	3810.3	5378.6	153.77	22.93	39.45	251
220	4.61476	1465	0.0473	3948.7	5574.0	154.67	22.92	38.71	255
225	4.46023	1533	0.0454	4084.4	5765.9	155.54	22.92	38.09	259
230	4.31833	1600	0.0436	4218.3	5955.1	156.37	22.93	37.57	263
235	4.18737	1665	0.0419	4350.7	6141.8	157.17	22.96	37.12	266
240	4.06595	1728	0.0405	4481.8	6326.4	157.95	22.99	36.74	270
245	3.95293	1790	0.0391	4611.9	6509.3	158.70	23.04	36.41	273
250	3.84735	1851	0.0378	4741.2	6690.6	159.43	23.08	36.13	276
255	3.74840	1911	0.0366	4869.8	6870.6	160.15	23.14	35.89	279
260	3.65539	1970	0.0356	4997.8	7049.5	160.84	23.20	35.68	282
265	3.56775	2028	0.0345	5125.3	7227.5	161.52	23.26	35.50	285
270	3.48497	2085	0.0336	5252.5	7404.6	162.18	23.33	35.35	288
275	3.40660	2142	0.0327	5379.4	7581.0	162.83	23.40	35.22	291
280	3.33225	2198	0.0318	5506.1	7756.8	163.46	23.47	35.11	294
285	3.26160	2253	0.0311	5632.6	7932.1	164.08	23.55	35.01	297
290	3.19434	2308	0.0303	5759.0	8106.9	164.69	23.63	34.93	300
295	3.13021	2362	0.0296	5885.5	8281.5	165.29	23.71	34.87	302
300	3.06898	2416	0.0289	6011.9	8455.7	165.87	23.79	34.82	305

* TWO-PHASE BOUNDARY

TABLE 16. THERMODYNAMIC PROPERTIES OF FLUORINE

8.0 MN/m² ISOBAR

TEMPERATURE (IPTS 1968) K	DENSITY MOL/L	ISOTHERM DERIVATIVE J/MOL	ISOCHORE DERIVATIVE MN/m²-K	INTERNAL ENERGY J/MOL	ENTHALPY J/MOL	ENTROPY J/MOL-K	C_v J / MOL - K	C_p J / MOL - K	VELOCITY OF SOUND M/S
* 54.259	45.0354	29405	4.3393	-6194.1	-6016.5	61.02	36.26	53.40	1067
56	44.7580	28327	4.3357	-6101.1	-5922.4	62.74	35.96	54.51	1063
58	44.4408	27338	4.3148	-5993.7	-5813.7	64.67	35.59	55.59	1060
60	44.1281	27053	4.2222	-5886.7	-5705.4	66.53	35.29	55.60	1059
62	43.8136	25871	4.0972	-5779.2	-5596.6	68.33	35.31	56.27	1042
64	43.4958	24272	3.9495	-5670.8	-5486.9	70.10	35.16	56.90	1017
66	43.1741	23436	3.8463	-5562.1	-5376.8	71.81	34.80	57.15	1006
68	42.8584	23412	3.7324	-5453.6	-5266.9	73.47	34.13	56.15	1007
70	42.5435	22624	3.6031	-5346.0	-5158.0	75.07	33.59	55.78	994
72	42.2201	20969	3.4818	-5237.5	-5048.0	76.63	33.26	56.61	969
74	41.8936	20032	3.3632	-5129.0	-4938.1	78.16	33.13	56.94	952
76	41.5624	19328	3.2471	-5019.4	-4826.9	79.66	32.82	56.82	938
78	41.2325	18600	3.1356	-4910.5	-4716.4	81.11	32.35	56.60	925
80	40.8981	17778	3.0254	-4801.4	-4605.8	82.52	32.08	56.71	909
82	40.5572	16626	2.9175	-4691.4	-4494.2	83.92	31.77	57.29	888
84	40.2164	16317	2.8184	-4581.7	-4382.8	85.27	31.38	56.67	881
86	39.8734	15596	2.7198	-4471.8	-4271.2	86.60	30.98	56.64	866
88	39.5273	14842	2.6192	-4361.6	-4159.2	87.90	30.54	56.57	851
90	39.1762	14234	2.5241	-4251.7	-4047.5	89.16	30.47	56.72	835
92	38.8200	13458	2.4282	-4140.9	-3934.9	90.41	30.52	57.27	815
94	38.4578	12724	2.3317	-4029.3	-3821.3	91.64	30.34	57.50	797
96	38.0910	12154	2.2432	-3917.3	-3707.3	92.85	30.02	57.41	782
98	37.7182	11507	2.1601	-3804.8	-3592.7	94.03	29.96	57.89	765
100	37.3415	10916	2.0774	-3691.6	-3477.3	95.20	29.79	58.13	749
102	36.9574	10280	1.9930	-3577.8	-3361.3	96.35	29.60	58.45	- 731
104	36.5668	9664	1.9190	-3463.0	-3244.2	97.48	29.49	59.13	714
106	36.1680	9177	1.8409	-3346.8	-3125.6	98.61	29.19	59.11	699
108	35.7639	8651	1.7659	-3230.1	-3006.4	99.72	28.92	59.35	684
110	35.3509	8121	1.6939	-3112.6	-2886.3	100.81	28.83	59.93	667
112	34.9280	7566	1.6232	-2993.5	-2764.5	101.90	28.72	60.69	649
114	34.4928	7065	1.5520	-2872.7	-2640.8	102.98	28.57	61.23	631
116	34.0479	6584	1.4847	-2750.6	-2515.6	104.05	28.40	61.91	615
118	33.5905	6123	1.4180	-2626.7	-2388.5	105.11	28.26	62.60	597
120	33.1201	5664	1.3523	-2500.9	-2259.4	106.17	28.11	63.42	580
122	32.6356	5216	1.2875	-2373.2	-2128.1	107.23	28.02	64.42	562
124	32.1326	4785	1.2239	-2243.0	-1994.0	108.29	27.97	65.57	543
126	31.6114	4361	1.1606	-2110.4	-1857.4	109.34	28.09	67.04	523
128	31.0665	3949	1.0983	-1973.7	-1716.2	110.41	27.98	68.49	504
130	30.5007	3569	1.0362	-1834.1	-1571.8	111.48	27.87	69.90	485
132	29.9048	3194	0.9729	-1690.0	-1422.5	112.56	27.53	71.26	467
134	29.2783	2835	0.9150	-1542.0	-1268.8	113.65	27.49	73.67	447
136	28.6145	2487	0.8555	-1248.8	-969.2	114.76	27.44	76.32	427
138	27.8988	2153	0.7955	-1100.0	-813.2	115.90	27.25	79.37	406
140	27.1315	1836	0.7350	-946.0	-651.1	117.07	27.17	83.14	385
142	26.2953	1536	0.6795	-784.5	-480.2	118.28	27.35	89.08	363
144	25.3633	1250	0.6170	-611.9	-296.5	119.57	27.52	95.70	338
146	24.3151	986	0.5559	-426.6	-97.6	120.94	27.68	105.05	314
148	23.1013	746	0.4910	-223.0	123.3	122.44	28.18	117.85	286
150	21.6605	536	0.4293	6.6	375.9	124.13	28.80	138.70	261
152	19.8936	371	0.3594	273.3	675.5	126.12	29.28	163.15	233
154	17.8008	266	0.2944	579.3	1028.8	128.43	29.72	187.79	210
156	15.5979	231	0.2430	904.3	1417.2	130.93	29.78	193.61	199
158	13.7253	249	0.2028	1199.4	1782.3	133.26	28.82	167.22	195
160	12.3220	292	0.1755	1438.4	2087.6	135.18	28.08	139.23	195
165	10.1513	426	0.1350	1867.3	2655.3	138.68	26.50	95.10	200
170	8.89257	554	0.1126	2171.8	3071.4	141.16	25.43	74.58	207
175	8.03802	672	0.0980	2417.1	3412.3	143.14	24.70	63.38	213
180	7.40156	780	0.0876	2628.9	3709.7	144.82	24.18	56.48	219
185	6.89938	880	0.0797	2819.5	3979.0	146.29	23.80	51.82	225
190	6.48823	973	0.0733	2995.4	4228.4	147.63	23.53	48.42	230
195	6.14150	1062	0.0682	3161.0	4463.6	148.85	23.33	45.95	235
200	5.84310	1147	0.0637	3318.8	4688.0	149.98	23.20	43.92	239
205	5.58253	1227	0.0600	3470.7	4903.7	151.05	23.11	42.44	243
210	5.35159	1303	0.0568	3617.9	5112.8	152.06	23.05	41.23	248
215	5.14477	1377	0.0540	3761.4	5316.4	153.02	23.01	40.24	252
220	4.95794	1449	0.0516	3901.9	5515.5	153.93	23.00	39.42	256
225	4.78793	1518	0.0493	4040.0	5710.8	154.81	22.99	38.73	259
230	4.63224	1586	0.0473	4176.0	5903.0	155.65	23.00	38.14	263
235	4.48889	1652	0.0455	4310.2	6092.4	156.47	23.02	37.65	267
240	4.35627	1717	0.0439	4443.1	6279.6	157.26	23.05	37.22	270
245	4.23305	1780	0.0423	4574.8	6464.7	158.02	23.09	36.85	273
250	4.11813	1842	0.0409	4705.5	6648.2	158.76	23.13	36.54	277
255	4.01060	1903	0.0396	4835.4	6830.1	159.48	23.18	36.26	280
260	3.90967	1963	0.0384	4964.7	7010.9	160.18	23.24	36.03	283
265	3.81467	2022	0.0373	5093.3	7190.5	160.87	23.30	35.83	286
270	3.72503	2080	0.0362	5221.6	7369.2	161.54	23.36	35.65	289
275	3.64026	2138	0.0353	5349.4	7547.1	162.19	23.43	35.50	292
280	3.55993	2194	0.0343	5477.0	7724.3	162.83	23.50	35.38	295
285	3.48364	2251	0.0335	5604.4	7900.9	163.45	23.58	35.18	298
290	3.41108	2306	0.0327	5731.7	8077.0	164.07	23.65	35.18	300
295	3.34195	2361	0.0319	5858.8	8252.7	164.67	23.73	35.10	303
300	3.27598	2415	0.0311	5986.0	8428.0	165.26	23.81	35.04	306

* TWO-PHASE BOUNDARY

TABLE 16. THERMODYNAMIC PROPERTIES OF FLUORINE

8.5 MN/m² ISOBAR

TEMPERATURE (IPTS 1968) K	DENSITY MOL/L	ISOTHERM DERIVATIVE J/MOL	ISOCHORE DERIVATIVE MN/m²-K	INTERNAL ENERGY J/MOL	ENTHALPY J/MOL	ENTROPY J/MOL-K	C_v J / MOL - K	C_p	VELOCITY OF SOUND M/S
* 54.306	45.0049	29524	4.3405	-6204.5	-6015.8	61.03	36.25	53.33	1069
56	44.7756	28446	4.3369	-6114.3	-5924.4	62.70	35.95	54.42	1065
58	44.4590	27423	4.3211	-6007.1	-5816.0	64.63	35.59	55.57	1062
60	44.1465	27133	4.2285	-5900.3	-5707.7	66.49	35.29	55.58	1060
62	43.8329	25936	4.1045	-5793.1	-5599.1	68.29	35.33	56.29	1043
64	43.5164	24319	3.9520	-5684.9	-5489.6	70.05	35.20	56.90	1017
66	43.1954	23484	3.8457	-5576.4	-5379.6	71.76	34.84	57.12	1007
68	42.8798	23469	3.7344	-5468.0	-5269.7	73.43	34.16	56.14	1007
70	42.5656	22704	3.6086	-5360.6	-5160.9	75.02	33.61	55.77	996
72	42.2439	21052	3.4875	-5252.4	-5051.2	76.59	33.28	56.59	971
74	41.9185	20101	3.3680	-5144.3	-4941.5	78.11	33.16	56.92	953
76	41.5883	19388	3.2500	-5034.9	-4830.5	79.61	32.87	56.81	939
78	41.2594	18677	3.1387	-4926.2	-4720.2	81.06	32.40	56.56	926
80	40.9262	17857	3.0293	-4817.4	-4609.7	82.47	32.12	56.67	911
82	40.5872	16712	2.9212	-4707.8	-4498.4	83.86	31.82	57.24	889
84	40.2469	16394	2.8216	-4598.3	-4387.1	85.22	31.43	56.62	882
86	39.9053	15685	2.7241	-4488.7	-4275.7	86.54	31.03	56.58	868
88	39.5609	14943	2.6247	-4378.8	-4164.0	87.84	30.58	56.50	852
90	39.2112	14336	2.5303	-4269.2	-4052.5	89.11	30.51	56.65	837
92	38.8570	13564	2.4355	-4158.9	-3940.2	90.35	30.56	57.21	817
94	38.4969	12830	2.3380	-4047.7	-3826.9	91.58	30.38	57.40	799
96	38.1320	12261	2.2501	-3936.0	-3713.1	92.78	30.06	57.32	784
98	37.7615	11614	2.1664	-3824.0	-3598.9	93.97	29.99	57.76	767
100	37.3871	11030	2.0846	-3711.2	-3483.9	95.13	29.83	58.02	751
102	37.0097	10394	1.9999	-3597.9	-3368.2	96.28	29.63	58.29	734
104	36.6183	9779	1.9264	-3483.7	-3251.6	97.41	29.53	58.96	717
106	36.2221	9292	1.8494	-3368.0	-3133.3	98.53	29.23	58.97	702
108	35.8213	8771	1.7742	-3251.9	-3014.6	99.64	28.94	59.15	687
110	35.4120	8244	1.7026	-3135.0	-2895.0	100.73	28.86	59.70	670
112	34.9935	7696	1.6327	-3016.7	-2773.8	101.81	28.74	60.42	653
114	34.5629	7185	1.5621	-2896.7	-2650.8	102.89	28.60	61.00	635
116	34.1231	6709	1.4951	-2775.4	-2526.3	103.95	28.42	61.62	619
118	33.6714	6248	1.4288	-2652.3	-2399.9	105.01	28.28	62.29	602
120	33.2074	5792	1.3636	-2527.6	-2271.6	106.07	28.12	63.06	585
122	32.7303	5350	1.2995	-2401.0	-2141.3	107.12	28.03	63.97	567
124	32.2357	4919	1.2363	-2272.1	-2008.4	108.17	27.98	65.06	549
126	31.7243	4498	1.1740	-2140.9	-1873.0	109.21	28.10	66.46	529
128	31.1909	4088	1.1122	-2005.8	-1733.3	110.27	27.98	67.80	511
130	30.6381	3711	1.0512	-1867.9	-1590.5	111.33	27.86	69.11	492
132	30.0579	3339	0.9880	-1725.9	-1443.1	112.40	27.52	70.24	474
134	29.4502	2982	0.9319	-1580.4	-1291.8	113.47	27.47	72.45	455
136	28.8096	2640	0.8731	-1274.7	-979.6	114.56	27.39	74.71	435
138	28.1231	2306	0.8149	-1129.4	-827.1	115.67	27.20	77.45	416
140	27.3927	1993	0.7561	-979.8	-669.5	116.81	27.08	80.61	395
142	26.6048	1697	0.6987	-824.0	-504.5	117.98	27.19	84.91	373
144	25.7392	1413	0.6427	-659.2	-329.0	119.20	27.36	90.91	352
146	24.7834	1152	0.5843	-484.5	-141.5	120.50	27.49	97.91	329
148	23.7066	911	0.5231	-296.6	62.0	121.88	27.79	106.89	304
150	22.4757	697	0.4669	-91.0	287.2	123.39	28.29	121.16	280
152	21.0305	518	0.4026	139.0	543.1	125.09	28.69	136.31	254
154	19.3626	385	0.3413	394.3	833.3	126.98	29.03	153.40	231
156	17.5018	304	0.2909	674.0	1159.7	129.09	29.31	170.89	216
158	15.6463	279	0.2442	957.2	1500.4	131.26	29.34	167.43	205
160	14.0501	292	0.2094	1215.1	1820.1	133.27	28.17	149.83	202
165	11.3697	398	0.1568	1701.9	2449.5	137.15	26.81	105.68	203
170	9.82304	522	0.1283	2042.8	2908.2	139.89	25.73	81.26	208
175	8.80079	640	0.1102	2310.5	3276.3	142.02	24.95	67.83	214
180	8.05570	750	0.0976	2537.3	3592.4	143.81	24.40	59.66	220
185	7.47702	852	0.0883	2738.7	3875.5	145.36	23.99	54.24	225
190	7.00898	947	0.0808	2922.9	4135.6	146.75	23.69	50.32	230
195	6.61782	1038	0.0749	3094.9	4379.3	148.01	23.47	47.51	235
200	6.28390	1124	0.0699	3257.9	4610.6	149.18	23.32	45.32	240
205	5.99350	1207	0.0655	3414.1	4832.3	150.28	23.21	43.50	244
210	5.73797	1285	0.0619	3565.0	5046.3	151.31	23.14	42.15	248
215	5.51006	1361	0.0587	3711.6	5254.2	152.29	23.10	41.05	252
220	5.30489	1434	0.0559	3854.9	5457.2	153.22	23.07	40.14	256
225	5.11876	1505	0.0535	3995.3	5655.9	154.12	23.07	39.37	260
230	4.94876	1573	0.0512	4133.5	5851.1	154.97	23.07	38.72	264
235	4.79260	1641	0.0492	4269.7	6043.3	155.80	23.08	38.17	267
240	4.64841	1706	0.0473	4404.3	6232.9	156.60	23.11	37.70	271
245	4.51469	1771	0.0457	4537.6	6420.4	157.37	23.14	37.29	274
250	4.39019	1834	0.0441	4669.8	6605.9	158.12	23.18	36.94	277
255	4.27385	1896	0.0427	4801.0	6789.9	158.85	23.23	36.64	281
260	4.16480	1957	0.0414	4931.5	6972.4	159.56	23.28	36.38	284
265	4.06228	2017	0.0401	5061.3	7153.7	160.25	23.34	36.15	287
270	3.96566	2076	0.0390	5190.6	7334.0	160.92	23.40	35.96	290
275	3.87437	2134	0.0379	5319.4	7513.3	161.58	23.46	35.79	293
280	3.78793	2192	0.0369	5448.0	7691.9	162.23	23.53	35.65	296
285	3.70592	2248	0.0359	5576.2	7869.8	162.86	23.60	35.52	298
290	3.62798	2305	0.0350	5704.3	8047.2	163.47	23.68	35.42	301
295	3.55376	2360	0.0342	5832.3	8224.0	164.08	23.75	35.33	304
300	3.48299	2415	0.0334	5960.1	8400.5	164.67	23.83	35.25	307

* TWO-PHASE BOUNDARY

TABLE 16. THERMODYNAMIC PROPERTIES OF FLUORINE

MN/m² ISOBAR

Σ	DENSITY MOL/L	ISOTHERM DERIVATIVE J/MOL	ISOCHORE DERIVATIVE MN/m²-K	INTERNAL ENERGY J/MOL	ENTHALPY J/MOL	ENTROPY J/MOL-K	C_v J/MOL-K	C_p J/MOL-K	VELOCITY OF SOUND M/S
♦	45.0542	29639	4.3417	-6214.9	-6015.1	61.04	36.24	53.27	1071
	44.7931	28569	4.3381	-6127.4	-5926.5	62.67	35.95	54.34	1066
	44.4772	27507	4.3273	-6020.5	-5818.2	64.59	35.58	55.54	1063
	44.1649	27213	4.2347	-5913.8	-5710.0	66.45	35.29	55.56	1062
	43.8521	26005	4.1123	-5806.9	-5601.6	68.25	35.34	56.30	1044
	43.5369	24375	3.9558	-5698.9	-5492.2	70.01	35.23	56.91	1018
	43.2166	23539	3.8452	-5590.7	-5382.4	71.72	34.89	57.08	1007
	42.9010	23528	3.7360	-5482.4	-5272.6	73.38	34.20	56.12	1008
	42.5876	22786	3.6136	-5375.2	-5163.9	74.98	33.63	55.75	997
	42.2676	21140	3.4931	-5267.4	-5054.4	76.54	33.30	56.56	972
	41.9433	20178	3.3731	-5159.5	-4944.9	78.06	33.19	56.91	954
	41.6140	19454	3.2531	-5050.3	-4834.0	79.56	32.92	56.79	940
	41.2861	18757	3.1416	-4941.8	-4723.8	81.01	32.44	56.52	927
	40.9541	17940	3.0333	-4833.4	-4613.6	82.42	32.16	56.63	912
	40.6170	16805	2.9251	-4724.1	-4502.6	83.81	31.87	57.18	891
	40.2774	16473	2.8246	-4614.9	-4391.4	85.17	31.48	56.56	883
	39.9371	15776	2.7282	-4505.6	-4280.2	86.49	31.07	56.51	869
	39.5943	15046	2.6300	-4396.0	-4168.7	87.79	30.62	56.43	854
	39.2459	14438	2.5363	-4286.7	-4057.4	89.05	30.54	56.57	839
	38.8938	13672	2.4428	-4176.8	-3945.4	90.29	30.60	57.14	820
	38.5357	12938	2.3456	-4066.0	-3832.4	91.52	30.41	57.33	801
	38.1726	12369	2.2570	-3954.7	-3718.9	92.72	30.09	57.23	787
	37.8043	11723	2.1729	-3843.1	-3605.1	93.90	30.02	57.64	-770
	37.4322	11142	2.0917	-3730.8	-3490.3	95.06	29.87	57.90	754
	37.0536	10508	2.0083	-3617.9	-3375.1	96.21	29.66	58.18	736
	36.6691	9896	1.9338	-3504.3	-3258.8	97.34	29.56	58.79	720
	36.2756	9406	1.8579	-3389.1	-3141.0	98.46	29.26	58.82	705
	35.8779	8890	1.7824	-3273.6	-3022.7	99.56	28.97	58.96	690
	35.4722	8366	1.7111	-3157.3	-2903.6	100.65	28.89	59.48	673
	35.0579	7826	1.6423	-3039.7	-2783.0	101.72	28.76	60.17	656
	34.6319	7306	1.5722	-2920.4	-2660.6	102.80	28.62	60.78	639
	34.1969	6833	1.5052	-2799.9	-2536.7	103.86	28.44	61.33	623
	33.7506	6371	1.4395	-2677.8	-2411.1	104.91	28.30	61.99	606
	33.2928	5920	1.3747	-2554.0	-2283.6	105.96	28.14	62.70	589
	32.8226	5484	1.3111	-2428.5	-2154.3	107.00	28.04	63.54	572
	32.3360	5052	1.2484	-2300.7	-2022.4	108.05	27.99	64.57	554
	31.8338	4635	1.1869	-2170.9	-1888.1	109.08	28.10	65.90	535
	31.3112	4226	1.1257	-2037.3	-1749.8	110.13	27.98	67.14	517
	30.7704	3851	1.0657	-1901.1	-1608.6	111.18	27.86	68.36	499
	30.2045	3481	1.0042	-1760.9	-1462.9	112.24	27.51	69.43	481
	29.6139	3128	0.9473	-1617.7	-1313.8	113.29	27.44	71.29	462
	28.9939	2789	0.8895	-1299.2	-988.7	114.36	27.35	73.24	443
	28.3332	2456	0.8335	-1157.0	-839.4	115.45	27.15	75.78	425
	27.6345	2145	0.7760	-1011.2	-685.5	116.56	27.02	78.48	405
	26.8866	1853	0.7200	-860.2	-525.5	117.70	27.07	82.01	384
	26.0745	1571	0.6664	-701.8	-356.6	118.68	27.22	87.09	364
	25.1897	1312	0.6097	-535.3	-178.0	120.11	27.33	92.55	342
	24.2123	1070	0.5515	-358.7	13.0	121.41	27.53	99.28	319
	23.1225	854	0.4992	-169.5	219.7	122.80	27.91	109.82	297
	21.8797	666	0.4391	37.4	448.7	124.31	26.23	120.16	273
	20.4820	516	0.3810	261.4	700.9	125.96	28.50	131.72	251
	18.9225	408	0.3332	504.4	980.0	127.76	28.80	147.35	234
	17.2698	346	0.2846	760.1	1281.3	129.68	28.98	153.03	219
	15.6841	326	0.2450	1011.1	1585.0	131.59	28.96	148.73	210
	12.6482	387	0.1810	1533.1	2244.7	135.65	27.01	114.27	208
	10.8045	499	0.1456	1909.6	2742.6	138.63	25.98	87.83	211
	9.59903	614	0.1235	2200.7	3138.3	140.92	25.18	72.39	216
	8.73484	724	0.1085	2443.4	3473.8	142.82	24.60	62.94	221
	8.07290	827	0.0974	2656.3	3771.1	144.45	24.16	56.73	226
	7.54341	924	0.0887	2849.1	4042.2	145.89	23.84	52.26	231
	7.10468	1016	0.0819	3027.9	4294.6	147.20	23.60	49.10	236
	6.73297	1104	0.0762	3196.3	4533.0	148.41	23.43	46.61	240
	6.41232	1187	0.0712	3356.9	4760.4	149.54	23.31	44.61	245
	6.13016	1267	0.0673	3511.6	4979.7	150.59	23.23	43.22	249
	5.87963	1345	0.0636	3661.5	5192.2	151.59	23.18	41.87	253
	5.65538	1420	0.0605	3807.6	5399.0	152.54	23.15	40.86	257
	5.45252	1492	0.0577	3950.5	5601.1	153.45	23.14	40.02	261
	5.26771	1562	0.0552	4090.9	5799.4	154.32	23.13	39.31	264
	5.09832	1630	0.0530	4229.1	5994.4	155.16	23.14	38.70	268
	4.94223	1697	0.0509	4365.5	6186.5	155.97	23.16	38.18	271
	4.79773	1763	0.0491	4500.4	6376.3	156.75	23.19	37.73	275
	4.66339	1827	0.0474	4634.0	6564.0	157.51	23.23	37.35	278
	4.53804	1889	0.0458	4766.6	6749.9	158.25	23.27	37.02	281
	4.42070	1951	0.0443	4898.3	6934.2	158.96	23.32	36.73	284
	4.31051	2012	0.0430	5029.3	7117.2	159.66	23.37	36.48	287
	4.20676	2072	0.0417	5159.6	7299.0	160.34	23.43	36.26	291
	4.10883	2131	0.0406	5289.5	7479.9	161.01	23.49	36.07	293
	4.01619	2189	0.0395	5418.9	7659.8	161.65	23.56	35.91	296
	3.92837	2247	0.0384	5548.0	7839.0	162.29	23.62	35.77	299
	3.84496	2304	0.0375	5676.9	8017.6	162.91	23.70	35.65	302
	3.76560	2360	0.0366	5805.5	8195.6	163.52	23.77	35.55	305
	3.68996	2416	0.0357	5934.1	8373.1	164.11	23.85	35.47	308

ASE BOUNDARY

TABLE 16. THERMODYNAMIC PROPERTIES OF FLUORINE

9.5 MN/m² ISOBAR

TEMPERATURE (IPTS 1968) K	DENSITY MOL/L	ISOTHERM DERIVATIVE J/MOL	ISOCHORE DERIVATIVE MN/m²-K	INTERNAL ENERGY J/MOL	ENTHALPY J/MOL	ENTROPY J/MOL-K	C_v J / MOL - K	C_p J / MOL - K	VELOCITY OF SOUND M/S
* 54.402	45.0634	29753	4.3428	-6225.2	-6014.4	61.05	36.23	53.21	1072
56	44.8106	28691	4.3392	-6140.5	-5928.5	62.63	35.95	54.25	1067
58	44.4954	27591	4.3335	-6033.9	-5820.4	64.55	35.58	55.52	1064
60	44.1833	27292	4.2408	-5927.4	-5712.3	66.41	35.28	55.54	1063
62	43.8713	26080	4.1207	-5820.6	-5604.1	68.21	35.34	56.32	1046
64	43.5574	24442	3.9602	-5713.0	-5494.9	69.97	35.26	56.91	1019
66	43.2378	23602	3.8448	-5604.9	-5385.2	71.68	34.93	57.04	1007
68	42.9223	23589	3.7371	-5496.7	-5275.4	73.34	34.24	56.10	1008
70	42.6095	22870	3.6181	-5389.8	-5166.8	74.94	33.65	55.72	998
72	42.2912	21234	3.4488	-5282.3	-5057.6	76.49	33.32	56.53	974
74	41.9680	20263	3.3785	-5174.6	-4948.3	78.01	33.21	56.88	956
76	41.6397	19526	3.2567	-5065.7	-4837.6	79.51	32.96	56.77	941
78	41.3127	18842	3.1443	-4957.5	-4727.5	80.96	32.49	56.47	928
80	40.9819	18028	3.0374	-4849.3	-4617.5	82.37	32.20	56.58	913
82	40.6467	16905	2.9290	-4740.4	-4506.7	83.76	31.92	57.11	892
84	40.3076	16555	2.8276	-4631.3	-4395.7	85.11	31.53	56.50	884
86	39.9687	15869	2.7322	-4522.3	-4284.7	86.44	31.12	56.44	870
88	39.6274	15151	2.6351	-4413.1	-4173.4	87.73	30.67	56.35	856
90	39.2805	14540	2.5419	-4304.1	-4062.3	88.99	30.57	56.49	841
92	38.9302	13781	2.4502	-4194.6	-3950.6	90.23	30.63	57.07	822
94	38.5742	13049	2.3534	-4084.2	-3837.9	91.46	30.44	57.26	804
96	38.2128	12477	2.2642	-3973.3	-3724.6	92.66	30.12	57.14	789
98	37.8468	11832	2.1795	-3862.1	-3611.1	93.84	30.05	57.52	772
100	37.4768	11254	2.0988	-3750.2	-3496.7	95.00	29.91	57.78	756
102	37.1009	10623	2.0166	-3637.9	-3381.8	96.14	29.70	58.06	739
104	36.7193	10014	1.9414	-3524.7	-3266.0	97.27	29.58	58.62	723
106	36.3284	9520	1.8665	-3410.0	-3148.5	98.38	29.29	58.68	708
108	35.9338	9009	1.7908	-3295.0	-3030.7	99.48	28.99	58.77	693
110	35.5315	8488	1.7199	-3179.4	-2912.0	100.57	28.91	59.28	677
112	35.1213	7955	1.6517	-3062.5	-2792.0	101.64	28.78	59.92	660
114	34.6998	7426	1.5823	-2944.0	-2670.2	102.71	28.64	60.56	643
116	34.2695	6957	1.5151	-2824.2	-2547.0	103.76	28.46	61.06	627
118	33.8284	6494	1.4500	-2702.9	-2422.1	104.81	28.32	61.70	610
120	33.3764	6046	1.3855	-2580.0	-2295.4	105.86	28.16	62.36	594
122	32.9127	5616	1.3225	-2455.5	-2166.9	106.89	28.06	63.13	577
124	32.4337	5184	1.2601	-2328.9	-2036.0	107.93	28.00	64.11	559
126	31.9401	4770	1.1994	-2200.3	-1902.9	108.96	28.11	65.36	540
128	31.4277	4362	1.1388	-2068.2	-1765.9	110.00	27.99	66.51	522
130	30.8979	3989	1.0796	-1933.5	-1626.0	111.04	27.86	67.65	505
132	30.3454	3621	1.0196	-1795.1	-1482.0	112.08	27.51	68.67	488
134	29.7702	3270	0.9618	-1653.9	-1334.8	113.12	27.42	70.18	469
136	29.1686	2935	0.9066	-1322.5	-996.8	114.18	27.31	72.07	452
138	28.5309	2603	0.8511	-1183.2	-856.2	115.25	27.10	74.28	433
140	27.8598	2294	0.7949	-1040.7	-712.3	116.33	26.96	76.64	414
142	27.1458	2006	0.7400	-893.7	-543.8	117.44	26.97	79.59	395
144	26.3781	1724	0.6883	-740.6	-380.5	118.58	27.10	83.95	375
146	25.5500	1466	0.6331	-580.7	-208.9	119.76	27.20	88.34	354
148	24.6486	1224	0.5773	-412.9	-27.5	121.00	27.33	93.64	332
150	23.6611	1006	0.5279	-235.6	165.9	122.30	27.63	101.81	312
152	22.5581	813	0.4709	-44.8	376.3	123.69	27.90	109.38	290
154	21.3418	652	0.4156	158.5	603.6	125.17	28.09	117.59	268
156	20.0006	526	0.3700	375.7	850.7	126.77	28.32	129.81	252
158	18.5566	438	0.3217	605.5	1117.5	128.47	28.55	136.94	235
160	17.0944	369	0.2799	839.0	1394.7	130.21	28.65	138.89	223
165	13.9297	396	0.2070	1368.3	2050.3	134.25	27.08	119.02	214
170	11.8211	487	0.1644	1774.3	2578.0	137.40	26.17	93.65	214
175	10.4270	595	0.1379	2088.6	2999.7	139.85	25.39	76.84	218
180	9.43619	703	0.1201	2347.7	3354.5	141.85	24.78	66.24	222
185	8.68538	806	0.1072	2572.5	3666.2	143.56	24.33	59.25	227
190	8.09043	904	0.0971	2774.2	3948.4	145.06	23.98	54.24	232
195	7.60131	998	0.0893	2960.1	4209.9	146.42	23.73	50.71	237
200	7.18969	1086	0.0827	3134.1	4455.5	147.66	23.54	47.93	241
205	6.83645	1171	0.0772	3299.3	4688.9	148.82	23.41	45.73	245
210	6.52703	1253	0.0728	3457.9	4913.4	149.90	23.32	44.18	250
215	6.25403	1331	0.0687	3611.1	5130.1	150.92	23.26	42.73	254
220	6.00936	1407	0.0653	3760.1	5341.0	151.89	23.23	41.71	258
225	5.78897	1481	0.0621	3905.6	5546.6	152.81	23.21	40.67	261
230	5.58888	1552	0.0593	4048.2	5748.0	153.70	23.20	39.89	265
235	5.40588	1621	0.0568	4188.4	5945.7	154.55	23.20	39.23	269
240	5.23757	1689	0.0546	4326.6	6140.4	155.37	23.22	38.66	272
245	5.08201	1755	0.0526	4463.1	6332.5	156.16	23.24	38.17	275
250	4.93761	1820	0.0507	4598.3	6522.3	156.93	23.27	37.75	279
255	4.80306	1884	0.0490	4732.2	6710.1	157.67	23.31	37.39	282
260	4.67725	1947	0.0474	4865.1	6896.2	158.40	23.35	37.07	285
265	4.55925	2008	0.0459	4997.2	7080.9	159.10	23.40	36.80	288
270	4.44825	2069	0.0446	5128.6	7264.3	159.79	23.46	36.56	291
275	4.34358	2129	0.0433	5259.5	7446.6	160.45	23.52	36.36	294
280	4.24465	2188	0.0421	5389.8	7627.4	161.11	23.58	36.18	297
285	4.15093	2246	0.0410	5519.8	7808.4	161.75	23.64	36.02	300
290	4.06198	2304	0.0400	5649.4	7988.2	162.37	23.71	35.89	303
295	3.97740	2361	0.0390	5778.8	8167.3	162.98	23.79	35.78	306
300	3.89684	2418	0.0380	5908.1	8346.0	163.58	23.86	35.68	308

* TWO-PHASE BOUNDARY

TABLE 16. THERMODYNAMIC PROPERTIES OF FLUORINE

10 MN/m² ISOBAR

PERATURE TS 1968) K	DENSITY MOL/L	ISOTHERM DERIVATIVE J/MOL	ISOCHORE DERIVATIVE MN/m²-K	INTERNAL ENERGY J/MOL	ENTHALPY J/MOL	ENTROPY J/MOL-K	C_v J / MOL - K	C_p	VELOCITY OF SOUND M/S
54.450	45.0727	29864	4.3439	-6235.6	-6013.7	61.06	36.22	53.15	1074
56	44.8280	28812	4.3403	-6153.6	-5930.5	62.59	35.94	54.16	1069
58	44.5135	27675	4.3358	-6047.3	-5822.6	64.51	35.58	55.46	1066
60	44.2016	27372	4.2469	-5940.9	-5714.6	66.37	35.28	55.51	1065
62	43.8905	26159	4.1295	-5834.4	-5606.6	68.17	35.35	56.33	1047
64	43.5778	24518	3.9651	-5727.0	-5497.5	69.92	35.30	56.91	1020
66	43.2590	23672	3.8445	-5619.1	-5388.0	71.63	34.97	56.99	1008
68	42.9434	23652	3.7379	-5511.0	-5276.2	73.30	34.28	56.07	1009
70	42.6313	22956	3.6222	-5404.3	-5169.7	74.89	33.67	55.69	1000
72	42.3147	21333	3.5044	-5297.2	-5060.8	76.45	33.34	56.49	975
74	41.9926	20357	3.3842	-5189.8	-4951.6	77.97	33.24	56.85	957
76	41.6652	19603	3.2605	-5081.1	-4841.1	79.46	33.00	56.75	942
78	41.3392	18930	3.1467	-4973.1	-4731.1	80.91	32.53	56.41	929
80	41.0096	18119	3.0415	-4865.2	-4621.3	82.32	32.24	56.53	914
82	40.6762	17012	2.9331	-4756.7	-4510.8	83.71	31.97	57.03	894
84	40.3378	16639	2.8304	-4647.8	-4399.9	85.06	31.58	56.44	885
86	40.0001	15964	2.7360	-4539.1	-4289.1	86.38	31.17	56.37	872
88	39.6603	15259	2.6401	-4430.2	-4178.0	87.68	30.71	56.26	858
90	39.3147	14642	2.5473	-4321.5	-4067.1	88.94	30.61	56.41	843
92	38.9663	13891	2.4575	-4212.4	-3955.8	90.17	30.66	57.00	824
94	38.6124	13161	2.3613	-4102.3	-3843.3	91.40	30.48	57.19	806
96	38.2527	12586	2.2714	-3991.8	-3730.3	92.60	30.16	57.05	792
98	37.8889	11943	2.1864	-3881.0	-3617.1	93.77	30.08	57.41	774
100	37.5210	11367	2.1058	-3769.5	-3503.0	94.93	29.95	57.66	759
102	37.1477	10739	2.0249	-3657.7	-3388.6	96.07	29.73	57.95	742
104	36.7690	10133	1.9492	-3545.1	-3273.1	97.19	29.61	58.45	726
106	36.3807	9634	1.8750	-3430.8	-3156.0	98.31	29.32	58.55	712
108	35.9889	9127	1.7992	-3316.4	-3038.5	99.41	29.02	58.59	696
110	35.5900	8608	1.7289	-3201.3	-2920.4	100.49	28.93	59.09	680
112	35.1836	8083	1.6610	-3085.1	-2800.8	101.56	28.80	59.68	664
114	34.7666	7547	1.5924	-2967.3	-2679.7	102.62	28.66	60.35	647
116	34.3407	7080	1.5249	-2848.3	-2557.1	103.67	28.48	60.78	631
118	33.9046	6617	1.4604	-2727.8	-2432.8	104.72	28.33	61.42	614
120	33.4582	6172	1.3961	-2605.8	-2306.9	105.76	28.17	62.02	598
122	33.0007	5746	1.3336	-2482.2	-2179.2	106.79	28.07	62.74	581
124	32.5289	5314	1.2716	-2356.7	-2049.3	107.81	28.01	63.67	564
126	32.0435	4903	1.2117	-2229.3	-1917.3	108.84	28.12	64.86	546
128	31.5406	4497	1.1517	-2098.5	-1781.5	109.87	27.99	65.94	528
130	31.0212	4125	1.0931	-1965.3	-1642.9	110.90	27.86	66.99	511
132	30.4809	3759	1.0344	-1828.5	-1500.4	111.93	27.50	67.94	494
134	29.9199	3411	0.9756	-1689.2	-1355.0	112.96	27.39	69.16	476
136	29.3349	3079	0.9228	-1344.7	-1003.8	114.00	27.28	70.99	459
138	28.7179	2747	0.8678	-1208.0	-859.8	115.05	27.05	72.92	441
140	28.0711	2440	0.8130	-1068.4	-712.2	116.11	26.92	75.03	423
142	27.3863	2154	0.7588	-925.0	-559.8	117.20	26.88	77.49	404
144	26.6562	1874	0.7068	-776.4	-401.3	118.30	26.99	81.01	385
146	25.8746	1616	0.6547	-622.0	-235.5	119.45	27.08	84.93	365
148	25.0337	1375	0.6012	-461.1	-61.7	120.63	27.17	89.27	345
150	24.1243	1156	0.5937	-292.9	121.6	121.86	27.40	95.79	326
152	23.1241	958	0.4993	-114.1	318.3	123.16	27.64	101.65	304
154	22.0374	790	0.4463	74.3	528.0	124.53	27.77	107.76	284
156	20.8539	651	0.4025	273.3	752.8	125.98	27.92	117.21	268
158	19.5802	544	0.3551	482.6	993.3	127.52	28.14	123.59	251
160	18.2618	473	0.3127	697.8	1245.4	129.10	28.28	127.47	237
165	15.1513	426	0.2339	1214.6	1874.6	132.97	27.03	119.38	222
170	12.8492	488	0.1844	1640.2	2418.5	135.22	26.31	98.09	219
175	11.2755	585	0.1534	1975.6	2862.5	138.80	25.56	80.92	221
180	10.1555	688	0.1324	2250.8	3235.5	140.90	24.95	69.44	225
185	9.31210	790	0.1175	2487.6	3561.5	142.69	24.48	61.75	229
190	8.64851	888	0.1059	2698.5	3854.8	144.25	24.12	56.22	233
195	8.10665	982	0.0971	2891.6	4125.1	145.66	23.85	52.32	238
200	7.65326	1071	0.0896	3071.5	4378.1	146.94	23.65	49.26	242
205	7.26606	1157	0.0834	3241.4	4617.7	148.12	23.50	46.86	246
210	6.92832	1240	0.0785	3404.0	4847.3	149.23	23.40	45.15	251
215	6.63141	1319	0.0739	3560.5	5068.5	150.27	23.33	43.57	255
220	6.36613	1396	0.0702	3712.5	5283.3	151.26	23.29	42.45	259
225	6.12812	1471	0.0666	3860.5	5492.3	152.20	23.27	41.35	262
230	5.91205	1543	0.0635	4005.4	5696.9	153.10	23.26	40.47	266
235	5.71508	1613	0.0608	4147.7	5897.4	153.96	23.26	39.76	269
240	5.53425	1682	0.0584	4287.7	6094.6	154.79	23.27	39.14	273
245	5.36739	1749	0.0561	4425.9	6289.0	155.59	23.29	38.61	276
250	5.21272	1815	0.0541	4562.5	6480.9	156.37	23.31	38.16	280
255	5.06879	1879	0.0522	4697.8	6670.6	157.12	23.35	37.76	283
260	4.93436	1943	0.0505	4832.0	6858.6	157.85	23.39	37.42	286
265	4.80841	2005	0.0489	4965.2	7044.9	158.56	23.43	37.12	289
270	4.69005	2067	0.0475	5097.7	7229.8	159.25	23.48	36.86	292
275	4.57854	2127	0.0461	5229.5	7413.6	159.93	23.54	36.64	295
280	4.47322	2187	0.0448	5360.7	7596.3	160.58	23.60	36.44	298
285	4.37352	2246	0.0436	5491.5	7778.0	161.23	23.66	36.27	301
290	4.27896	2305	0.0425	5622.0	7959.0	161.86	23.73	36.12	304
295	4.18911	2362	0.0414	5752.1	8139.3	162.47	23.80	36.00	307
300	4.10357	2420	0.0404	5882.1	8319.0	163.08	23.87	35.89	309

TWO-PHASE BOUNDARY

TABLE 16. THERMODYNAMIC PROPERTIES OF FLUORINE

11 MN/m² ISOBAR

TEMPERATURE (IPTS 1968) K	DENSITY MOL/L	ISOTHERM DERIVATIVE J/MOL	ISOCHORE DERIVATIVE MN/m²-K	INTERNAL ENERGY J/MOL	ENTHALPY J/MOL	ENTROPY J/MOL-K	C_v J / MOL	C_p - K	VELOCITY OF SOUND M/S
* 54.546	45.0911	30082	4.3458	-6256.3	-6012.3	61.09	36.19	53.04	1077
56	44.8626	29054	4.3423	-6179.7	-5934.5	62.52	35.94	53.99	1072
58	44.5495	27842	4.3376	-6073.9	-5827.0	64.43	35.57	55.32	1067
60	44.2380	27530	4.2589	-5967.8	-5719.2	66.29	35.27	55.47	1067
62	43.9286	26333	4.1486	-5861.9	-5611.5	68.08	35.36	56.36	1051
64	43.6185	24698	3.9767	-5754.9	-5502.8	69.84	35.36	56.89	1023
66	43.3011	23833	3.8443	-5647.5	-5393.5	71.55	35.05	56.88	1009
68	42.9856	23785	3.7382	-5539.6	-5283.7	73.21	34.37	55.99	1010
70	42.6747	23132	3.6290	-5433.2	-5175.5	74.81	33.72	55.61	1002
72	42.3613	21546	3.5157	-5326.8	-5067.1	76.35	33.38	56.40	979
74	42.0415	20570	3.3957	-5219.9	-4958.3	77.87	33.28	56.75	961
76	41.7160	19776	3.2691	-5111.8	-4848.1	79.37	33.08	56.69	944
78	41.3917	19119	3.1510	-5004.1	-4738.4	80.82	32.62	56.27	932
80	41.0645	18312	3.0490	-4896.8	-4628.9	82.22	32.32	56.40	917
82	40.7346	17246	2.9414	-4789.0	-4519.0	83.60	32.06	56.85	897
84	40.3976	16814	2.8357	-4680.5	-4408.2	84.96	31.68	56.29	887
86	40.0624	16159	2.7431	-4572.4	-4297.8	86.28	31.26	56.21	874
88	39.7253	15481	2.6494	-4464.1	-4187.2	87.57	30.79	56.08	861
90	39.3826	14844	2.5573	-4356.0	-4076.7	88.82	30.68	56.24	846
92	39.0377	14116	2.4711	-4247.7	-3965.9	90.06	30.71	56.83	829
94	38.6877	13392	2.3773	-4138.4	-3854.0	91.28	30.54	57.04	811
96	38.3315	12806	2.2863	-4028.5	-3741.5	92.47	30.21	56.88	797
98	37.9718	12166	2.2009	-3918.6	-3628.9	93.65	30.12	57.19	780
100	37.6081	11593	2.1202	-3807.9	-3515.4	94.80	30.01	57.42	764
102	37.2398	10973	2.0412	-3696.9	-3401.6	95.94	29.78	57.71	748
104	36.8665	10375	1.9650	-3585.4	-3287.0	97.05	29.64	58.12	732
106	36.4832	9861	1.8920	-3472.1	-3170.6	98.16	29.37	58.27	718
108	36.0971	9363	1.8179	-3358.6	-3053.9	99.26	29.05	58.30	703
110	35.7046	8848	1.7471	-3244.8	-2936.7	100.33	28.96	58.73	687
112	35.3054	8338	1.6792	-3129.7	-2818.1	101.39	28.83	59.22	671
114	34.8970	7790	1.6124	-3013.4	-2698.2	102.45	28.69	59.94	654
116	34.4795	7326	1.5441	-2895.7	-2576.7	103.49	28.51	60.26	638
118	34.0530	6861	1.4810	-2776.8	-2453.8	104.53	28.35	60.89	623
120	33.6171	6421	1.4177	-2656.5	-2329.3	105.56	28.19	61.43	607
122	33.1709	6005	1.3560	-2534.7	-2203.1	106.58	28.09	62.04	591
124	32.7127	5573	1.2952	-2411.2	-2074.9	107.59	28.02	62.90	574
126	32.2421	5167	1.2357	-2286.0	-1944.9	108.60	28.13	63.95	556
128	31.7566	4764	1.1771	-2157.7	-1811.3	109.62	27.99	64.91	539
130	31.2560	4394	1.1195	-2027.1	-1675.2	110.63	27.86	65.81	523
132	30.7377	4031	1.0624	-1893.4	-1535.5	111.64	27.50	66.62	507
134	30.2018	3686	1.0053	-1757.3	-1393.1	112.65	27.35	67.62	490
136	29.6456	3359	0.9523	-1386.6	-1015.5	113.67	27.23	69.00	473
138	29.0644	3029	0.8994	-1254.3	-875.8	114.69	26.96	70.59	457
140	28.4585	2725	0.8473	-1119.7	-733.2	115.71	26.84	72.37	440
142	27.8219	2442	0.7945	-982.0	-586.6	116.75	26.76	74.19	422
144	27.1521	2164	0.7440	-840.8	-435.6	117.81	26.82	76.79	404
146	26.4435	1905	0.6941	-695.0	-279.1	118.89	26.91	79.71	385
148	25.6934	1664	0.6444	-544.8	-116.7	119.99	26.95	82.90	367
150	24.8962	1444	0.5994	-389.6	52.2	121.13	27.05	87.25	350
152	24.0386	1240	0.5489	-227.6	230.0	122.30	27.25	91.16	330
154	23.1254	1061	0.4994	-59.3	416.4	123.52	27.35	95.01	311
156	22.1503	907	0.4580	115.6	612.2	124.78	27.37	100.94	297
158	21.1127	777	0.4130	297.3	818.3	126.10	27.45	105.24	280
160	20.0302	675	0.3711	484.0	1033.1	127.45	27.59	108.90	265
165	17.2601	537	0.2869	954.4	1591.7	130.89	26.64	111.52	243
170	14.8313	531	0.2269	1388.1	2129.8	134.10	26.38	101.36	232
175	12.9850	592	0.1867	1753.0	2600.1	136.83	25.80	86.96	229
180	11.6239	678	0.1593	2056.8	3003.1	139.10	25.23	75.07	231
185	10.5950	772	0.1398	2326.7	3354.9	141.03	24.74	66.43	234
190	9.78972	867	0.1250	2549.9	3669.5	142.71	24.36	60.47	237
195	9.13804	960	0.1136	2753.5	3957.2	144.20	24.07	55.47	241
200	8.59705	1050	0.1043	2945.1	4224.6	145.56	23.84	51.90	245
205	8.13875	1136	0.0966	3124.8	4478.3	146.80	23.67	49.09	249
210	7.74183	1220	0.0905	3295.5	4716.4	147.96	23.55	47.08	253
215	7.39508	1301	0.0849	3459.0	4946.5	149.04	23.47	45.23	257
220	7.08696	1380	0.0804	3616.9	5169.0	150.06	23.41	43.92	261
225	6.81183	1456	0.0761	3770.1	5385.0	151.03	23.38	42.65	264
230	6.56294	1530	0.0726	3919.7	5595.8	151.96	23.36	41.75	268
235	6.33735	1602	0.0691	4066.1	5801.8	152.85	23.36	40.80	271
240	6.13058	1672	0.0663	4209.9	6004.2	153.70	23.36	40.15	275
245	5.94082	1739	0.0635	4351.4	6203.0	154.52	23.37	39.48	278
250	5.76506	1807	0.0611	4491.0	6399.1	155.31	23.39	38.95	281
255	5.60189	1873	0.0590	4629.1	6592.7	156.08	23.42	38.50	285
260	5.44982	1938	0.0569	4765.7	6784.1	156.82	23.45	38.10	288
265	5.30762	2002	0.0551	4901.3	6973.8	157.55	23.48	37.75	291
270	5.17423	2065	0.0534	5035.8	7161.7	158.25	23.53	37.45	294
275	5.04875	2127	0.0518	5169.6	7348.3	158.93	23.58	37.19	297
280	4.93042	2188	0.0503	5302.6	7533.7	159.60	23.63	36.96	300
285	4.81857	2248	0.0489	5435.1	7717.9	160.25	23.68	36.75	303
290	4.71261	2308	0.0476	5567.1	7901.3	160.89	23.75	36.58	306
295	4.61204	2367	0.0464	5698.7	8083.8	161.51	23.81	36.43	309
300	4.51640	2425	0.0453	5830.0	8265.6	162.13	23.88	36.30	312

* TWO-PHASE BOUNDARY

TABLE 16. THERMODYNAMIC PROPERTIES OF FLUORINE

12 MN/m² ISOBAR

TEMPERATURE (IPTS 1968) K	DENSITY MOL/L	ISOTHERM DERIVATIVE J/MOL	ISOCHORE DERIVATIVE MN/m²-K	INTERNAL ENERGY J/MOL	ENTHALPY J/MOL	ENTROPY J/MOL-K	C_v J / MOL - K	C_p J / MOL - K	VELOCITY OF SOUND M/S
* 54.641	45.1093	30291	4.3474	-6276.9	-6010.9	61.11	36.17	52.93	1080
56	44.8968	29293	4.3441	-6205.7	-5938.5	62.44	35.93	53.83	1075
58	44.5853	28008	4.3392	-6100.5	-5831.3	64.36	35.57	55.18	1069
60	44.2742	27687	4.2707	-5994.7	-5723.7	66.21	35.27	55.43	1070
62	43.9664	26526	4.1695	-5889.2	-5616.3	68.00	35.37	56.39	1055
64	43.6588	24915	3.9904	-5782.8	-5507.9	69.75	35.41	56.87	1026
66	43.3429	24022	3.8445	-5675.7	-5398.9	71.46	35.13	56.75	1011
68	43.0275	23925	3.7371	-5568.1	-5289.2	73.13	34.46	55.90	1011
70	42.7177	23315	3.6341	-5462.1	-5181.2	74.72	33.78	55.51	1004
72	42.4075	21779	3.5269	-5356.3	-5073.3	76.27	33.42	56.29	983
74	42.0899	20815	3.4072	-5250.0	-4964.9	77.78	33.33	56.63	965
76	41.7663	19970	3.2790	-5142.3	-4855.0	79.27	33.16	56.62	947
78	41.4437	19323	3.1544	-5035.0	-4745.4	80.72	32.72	56.10	934
80	41.1188	18520	3.0556	-4928.2	-4636.4	82.13	32.40	56.25	920
82	40.7922	17505	2.9500	-4821.1	-4526.9	83.50	32.15	56.65	901
84	40.4567	16997	2.8406	-4713.0	-4416.4	84.86	31.78	56.14	889
86	40.1239	16361	2.7496	-4605.5	-4306.4	86.17	31.35	56.04	877
88	39.7895	15711	2.6580	-4497.8	-4196.2	87.46	30.88	55.87	865
90	39.4495	15046	2.5663	-4390.3	-4086.1	88.71	30.75	56.06	850
92	39.1080	14344	2.4830	-4282.7	-3975.9	89.94	30.77	56.62	833
94	38.7617	13631	2.3930	-4174.1	-3864.5	91.16	30.60	56.89	817
96	38.4089	13028	2.3015	-4064.9	-3752.5	92.35	30.25	56.71	802
98	38.0533	12391	2.2171	-3955.8	-3640.5	93.52	30.16	57.01	785
100	37.6936	11821	2.1355	-3845.9	-3527.6	94.67	30.05	57.20	770
102	37.3300	11211	2.0569	-3735.8	-3414.4	95.80	29.83	57.46	754
104	36.9617	10622	1.9816	-3625.2	-3300.5	96.91	29.66	57.81	738
106	36.5835	10088	1.9090	-3512.9	-3184.9	98.02	29.40	58.01	724
108	36.2026	9596	1.8366	-3400.4	-3068.9	99.11	29.07	58.04	710
110	35.8161	9085	1.7656	-3287.6	-2952.5	100.18	28.98	58.40	694
112	35.4236	8589	1.6971	-3173.7	-2834.9	101.23	28.85	58.78	679
114	35.0234	8036	1.6319	-3058.7	-2716.1	102.28	28.72	59.52	662
116	34.6138	7572	1.5648	-2942.4	-2595.7	103.32	28.53	59.84	647
118	34.1963	7105	1.5014	-2824.9	-2474.0	104.34	28.37	60.39	631
120	33.7699	6669	1.4392	-2706.1	-2350.8	105.36	28.21	60.89	615
122	33.3340	6259	1.3775	-2586.0	-2226.0	106.38	28.10	61.39	600
124	32.8881	5829	1.3176	-2464.4	-2099.5	107.38	28.04	62.18	583
126	32.4309	5428	1.2586	-2341.2	-1971.2	108.38	28.14	63.10	566
128	31.9608	5028	1.2009	-2215.1	-1839.6	109.38	28.00	63.95	550
130	31.4770	4658	1.1443	-2086.9	-1705.7	110.37	27.86	64.74	534
132	30.9779	4297	1.0885	-1955.8	-1568.4	111.37	27.49	65.42	519
134	30.4636	3955	1.0333	-1822.5	-1428.6	112.36	27.33	66.31	503
136	29.9318	3631	0.9794	-1425.2	-102.3	113.35	27.17	67.27	486
138	29.3804	3302	0.9295	-1296.7	-888.3	114.35	26.89	68.71	471
140	28.8079	3001	0.8787	-1166.3	-749.7	115.34	26.76	70.16	455
142	28.2097	2719	0.8276	-1033.2	-607.8	116.35	26.68	71.64	438
144	27.5867	2442	0.7779	-897.6	-462.6	117.37	26.69	73.57	421
146	26.9333	2183	0.7303	-758.6	-313.1	118.40	26.78	75.96	404
148	26.2499	1942	0.6831	-616.2	-159.0	119.45	26.80	78.42	387
150	25.5293	1721	0.6390	-470.1	-0.0	120.51	26.83	81.43	371
152	24.7671	1513	0.5915	-319.3	165.2	121.61	26.93	84.24	353
154	23.9657	1327	0.5447	-164.1	336.6	122.73	27.06	86.99	335
156	23.1215	1163	0.5049	-4.4	514.6	123.88	27.06	91.04	321
158	22.2335	1019	0.4619	159.8	699.5	125.05	27.04	93.95	305
160	21.3108	899	0.4212	327.5	890.6	126.26	27.08	96.61	291
165	18.8976	697	0.3361	794.0	1389.0	129.32	26.18	101.06	266
170	16.5785	624	0.2700	1170.5	1894.3	132.34	26.19	98.45	248
175	14.6206	638	0.2222	1545.1	2365.9	135.07	25.87	89.22	241
180	13.0825	698	0.1884	1868.4	2785.7	137.44	25.40	78.91	239
185	11.8889	777	0.1640	2147.8	3157.1	139.48	24.95	70.25	240
190	10.9470	864	0.1457	2393.8	3490.0	141.25	24.56	63.49	242
195	10.1855	952	0.1315	2615.3	3793.4	142.83	24.25	58.38	246
200	9.55495	1040	0.1202	2818.6	4074.5	144.25	24.01	54.43	249
205	9.02343	1126	0.1107	3007.9	4337.8	145.55	23.83	51.25	252
210	8.56537	1210	0.1033	3186.9	4587.9	146.76	23.69	48.95	256
215	8.16703	1291	0.0965	3357.3	4826.6	147.88	23.59	46.85	260
220	7.81458	1370	0.0911	3521.3	5056.8	148.94	23.52	45.35	264
225	7.50108	1447	0.0860	3679.7	5279.5	149.94	23.48	43.92	267
230	7.21846	1522	0.0819	3834.0	5496.4	150.90	23.45	42.89	271
235	6.96302	1596	0.0778	3984.5	5707.8	151.81	23.44	41.84	274
240	6.72961	1667	0.0745	4132.0	5915.2	152.68	23.44	41.09	277
245	6.51569	1738	0.0714	4276.9	6118.6	153.52	23.45	40.36	281
250	6.31819	1806	0.0686	4419.7	6319.0	154.33	23.46	39.80	284
255	6.13549	1874	0.0660	4560.6	6516.4	155.11	23.48	39.23	287
260	5.96618	1936	0.0636	4699.7	6711.1	155.87	23.50	38.77	290
265	5.80733	2001	0.0615	4837.5	6903.9	156.60	23.53	38.37	293
270	5.65857	2065	0.0595	4974.4	7094.8	157.32	23.56	38.02	296
275	5.51883	2129	0.0577	5109.8	7284.2	158.01	23.60	37.72	299
280	5.38723	2191	0.0560	5244.6	7472.1	158.69	23.65	37.46	302
285	5.26298	2253	0.0544	5378.7	7658.8	159.35	23.70	37.23	305
290	5.14541	2314	0.0529	5512.3	7844.4	159.99	23.75	37.02	308
295	5.03394	2374	0.0516	5645.3	8029.1	160.63	23.81	36.85	311
300	4.92804	2434	0.0503	5777.9	8213.0	161.24	23.87	36.69	314

* TWO-PHASE BOUNDARY

TABLE 16. THERMODYNAMIC PROPERTIES OF FLUORINE

13 MN/m² ISOBAR

TEMPERATURE (IPTS 1968) K	DENSITY MOL/L	ISOTHERM DERIVATIVE J/MOL	ISOCHORE DERIVATIVE MN/m²-K	INTERNAL ENERGY J/MOL	ENTHALPY J/MOL	ENTROPY J/MOL-K	C_v J / MOL - K	C_p J / MOL - K	VELOCITY OF SOUND M/S
* 54.737	45.1275	30494	4.3488	-6297.6	-6009.5	61.13	36.15	52.82	1083
56	44.9308	29530	4.3457	-6231.7	-5942.3	62.37	35.93	53.67	1077
58	44.6209	28172	4.3407	-6126.9	-5835.6	64.28	35.56	55.04	1071
60	44.3102	27844	4.2822	-6021.6	-5728.2	66.13	35.26	55.39	1073
62	44.0040	26737	4.1909	-5916.5	-5621.1	67.92	35.37	56.40	1059
64	43.6987	25168	4.0061	-5810.6	-5513.1	69.67	35.46	56.83	1030
66	43.3843	24238	3.8450	-5703.8	-5404.2	71.38	35.21	56.60	1013
68	43.0692	24074	3.7363	-5596.5	-5294.7	73.04	34.54	55.80	1012
70	42.7605	23504	3.6374	-5490.8	-5186.8	74.64	33.84	55.39	1006
72	42.4531	22031	3.5381	-5385.7	-5079.5	76.18	33.46	56.16	986
74	42.1376	21090	3.4186	-5279.8	-4971.3	77.69	33.38	56.47	969
76	41.8161	20185	3.2900	-5172.6	-4861.8	79.18	33.23	56.54	951
78	41.4952	19541	3.1569	-5065.7	-4752.4	80.63	32.81	55.91	936
80	41.1725	18743	3.0612	-4959.5	-4643.7	82.03	32.48	56.07	923
82	40.8488	17789	2.9589	-4853.0	-4534.7	83.40	32.23	56.42	905
84	40.5152	17189	2.8461	-4745.4	-4424.5	84.75	31.87	55.99	891
86	40.1847	16569	2.7557	-4638.4	-4314.9	86.07	31.45	55.85	880
88	39.8526	15949	2.6659	-4531.2	-4205.0	87.35	30.96	55.65	869
90	39.5155	15247	2.5747	-4424.4	-4095.4	88.61	30.83	55.89	853
92	39.1772	14577	2.4937	-4317.4	-3985.6	89.83	30.83	56.40	838
94	38.8344	13877	2.4082	-4209.6	-3874.8	91.04	30.66	56.71	822
96	38.4850	13252	2.3170	-4101.1	-3763.3	92.23	30.28	56.54	807
98	38.1332	12620	2.2342	-3992.7	-3651.8	93.40	30.18	56.84	791
100	37.7773	12049	2.1517	-3883.6	-3539.5	94.55	30.08	57.01	775
102	37.4183	11451	2.0718	-3774.3	-3426.9	95.67	29.88	57.19	759
104	37.0548	10873	1.9992	-3664.6	-3313.8	96.78	29.68	57.52	745
106	36.6815	10313	1.9266	-3553.2	-3198.8	97.88	29.41	57.77	730
108	36.3055	9828	1.8551	-3441.6	-3083.5	98.96	29.09	57.79	717
110	35.9248	9320	1.7846	-3329.8	-2968.0	100.03	28.98	58.10	701
112	35.5384	8836	1.7154	-3217.0	-2851.2	101.08	28.87	58.40	686
114	35.1460	8284	1.6510	-3103.4	-2733.5	102.12	28.73	59.10	670
116	34.7438	7817	1.5854	-2988.2	-2614.1	103.15	28.54	59.44	655
118	34.3346	7349	1.5213	-2872.2	-2493.6	104.17	28.38	59.90	639
120	33.9171	6915	1.4600	-2754.8	-2371.5	105.18	28.22	60.38	624
122	33.4907	6509	1.3984	-2636.2	-2248.0	106.18	28.11	60.79	609
124	33.0560	6083	1.3394	-2516.3	-2123.0	107.17	28.04	61.51	593
126	32.6109	5684	1.2813	-2394.9	-1996.3	108.16	28.14	62.36	576
128	32.1548	5287	1.2243	-2270.9	-1866.6	109.15	28.01	63.10	560
130	31.6859	4918	1.1687	-2144.8	-1734.5	110.13	27.86	63.82	545
132	31.2038	4558	1.1139	-2016.0	-1599.4	111.11	27.48	64.39	530
134	30.7084	4218	1.0598	-1885.4	-1462.0	112.09	27.31	65.16	515
136	30.1977	3895	1.0072	-1461.2	-1030.7	113.06	27.10	65.95	499
138	29.6715	3570	0.9571	-1336.0	-897.9	114.03	26.84	67.06	485
140	29.1270	3270	0.9083	-1209.1	-762.8	115.00	26.69	68.32	469
142	28.5604	2987	0.8582	-1079.8	-624.6	115.98	26.63	69.56	453
144	27.9750	2712	0.8094	-948.8	-484.1	116.97	26.60	71.05	437
146	27.3653	2451	0.7632	-815.2	-340.1	117.96	26.67	73.00	420
148	26.7310	2211	0.7175	-678.7	-192.4	118.97	26.71	74.95	404
150	26.0691	1989	0.6744	-539.4	-40.8	119.98	26.70	77.17	389
152	25.3760	1778	0.6292	-396.9	115.4	121.02	26.74	79.31	373
154	24.6535	1587	0.5847	-251.1	276.2	122.07	26.81	81.38	356
156	23.8996	1415	0.5460	-101.9	442.1	123.14	26.83	84.35	342
158	23.1137	1262	0.5045	50.4	612.9	124.23	26.78	86.45	327
160	22.3016	1129	0.4650	205.1	788.0	125.33	26.77	88.37	313
165	20.1717	883	0.3806	597.4	1241.9	128.12	25.83	92.33	288
170	18.0384	755	0.3114	990.0	1710.6	130.92	25.97	93.04	267
175	16.0997	721	0.2582	1359.9	2167.4	133.57	25.80	88.20	255
180	14.4717	747	0.2189	1692.2	2590.5	135.95	25.45	80.57	249
185	13.1561	805	0.1897	1985.2	2973.3	138.05	25.06	72.62	248
190	12.0963	879	0.1677	2245.1	3319.8	139.90	24.70	66.22	249
195	11.2329	960	0.1505	2479.1	3636.4	141.54	24.39	60.88	251
200	10.5158	1043	0.1371	2693.3	3929.5	143.03	24.15	56.72	254
205	9.91192	1127	0.1258	2891.9	4203.5	144.38	23.96	53.24	257
210	9.39274	1209	0.1169	3078.8	4462.8	145.63	23.81	50.70	260
215	8.94241	1290	0.1088	3256.1	4709.8	146.80	23.70	48.39	263
220	8.54512	1369	0.1024	3426.0	4947.4	147.89	23.62	46.73	267
225	8.19270	1446	0.0964	3589.7	5176.5	148.92	23.57	45.14	270
230	7.87584	1521	0.0916	3748.5	5399.1	149.90	23.54	43.99	274
235	7.59002	1595	0.0869	3903.1	5615.9	150.83	23.52	42.85	277
240	7.32949	1668	0.0831	4054.4	5828.0	151.72	23.51	42.01	280
245	7.09113	1739	0.0795	4202.7	6035.9	152.58	23.52	41.22	283
250	6.87154	1809	0.0763	4348.5	6240.4	153.41	23.53	40.59	287
255	6.66889	1877	0.0733	4492.2	6441.5	154.20	23.54	39.94	289
260	6.48050	1943	0.0707	4634.0	6640.0	154.98	23.56	39.47	293
265	6.30509	2009	0.0681	4774.2	6836.0	155.72	23.58	38.98	296
270	6.14092	2074	0.0659	4912.9	7029.9	156.45	23.60	38.60	299
275	5.98814	2133	0.0638	5050.3	7221.3	157.15	23.62	38.24	302
280	5.84306	2197	0.0619	5186.9	7411.7	157.84	23.65	37.95	305
285	5.70624	2260	0.0601	5322.6	7600.8	158.51	23.70	37.68	308
290	5.57690	2322	0.0584	5457.6	7788.6	159.16	23.74	37.45	310
295	5.45438	2383	0.0569	5592.0	7975.4	159.80	23.80	37.25	313
300	5.33809	2444	0.0554	5725.8	8161.2	160.42	23.85	37.08	316

* TWO-PHASE BOUNDARY

TEMPERATURE (IPTS 1968) K	DENSITY MOL/L	ISOTHERM DERIVATIVE J/MOL	ISOCHORE DERIVATIVE MN/m²-K	INTERNAL ENERGY J/MOL	ENTHALPY J/MOL	ENTROPY J/MOL-K	C_v J / MOL - K	C_p J / MOL - K	VELOCITY OF SOUND M/S
* 54.832	45.1455	30689	4.3500	-6318.2	-6008.1	61.16	36.13	52.72	1086
56	44.9646	29766	4.3470	-6257.5	-5946.2	62.30	35.92	53.51	1080
58	44.6563	28336	4.3419	-6153.3	-5839.8	64.20	35.56	54.91	1073
60	44.3460	27999	4.2934	-6048.3	-5732.6	66.06	35.26	55.35	1075
62	44.0412	26967	4.2025	-5943.7	-5625.8	67.84	35.36	56.30	1063
64	43.7382	25454	4.0237	-5838.2	-5518.2	69.58	35.50	56.78	1035
66	43.4254	24480	3.8460	-5731.8	-5409.4	71.29	35.29	56.43	1015
68	43.1106	24230	3.7359	-5624.8	-5300.0	72.96	34.63	55.70	1013
70	42.8028	23699	3.6391	-5519.4	-5192.3	74.55	33.90	55.25	1008
72	42.4982	22301	3.5493	-5414.9	-5085.5	76.09	33.50	56.02	991
74	42.1847	21394	3.4300	-5309.5	-4977.7	77.60	33.42	56.29	974
76	41.8654	20421	3.3020	-5202.9	-4868.5	79.08	33.29	56.44	955
78	41.5461	19773	3.1651	-5096.3	-4759.3	80.53	32.90	55.80	939
80	41.2255	18978	3.0659	-4990.5	-4650.9	81.93	32.56	55.87	926
82	40.9046	18095	2.9681	-4884.7	-4542.4	83.30	32.31	56.17	910
84	40.5731	17389	2.8546	-4777.5	-4432.4	84.65	31.97	55.88	894
86	40.2446	16784	2.7613	-4671.0	-4323.1	85.97	31.54	55.66	883
88	39.9149	16195	2.6731	-4564.4	-4213.6	87.25	31.05	55.42	872
90	39.5806	15447	2.5840	-4458.2	-4104.5	88.50	30.91	55.74	856
92	39.2452	14813	2.5034	-4351.9	-3995.2	89.72	30.89	56.17	842
94	38.9058	14130	2.4224	-4244.7	-3884.9	90.93	30.72	56.51	827
96	38.5599	13478	2.3339	-4136.9	-3773.9	92.12	30.30	56.40	812
98	38.2118	12851	2.2514	-4029.3	-3662.9	93.28	30.20	56.67	797
100	37.8596	12278	2.1691	-3920.9	-3551.1	94.42	30.10	56.83	781
102	37.5047	11693	2.0859	-3812.4	-3439.1	95.54	29.92	56.90	765
104	37.1457	11127	2.0165	-3703.5	-3326.6	96.65	29.69	57.23	751
106	36.7775	10538	1.9449	-3593.1	-3212.5	97.74	29.42	57.55	737
108	36.4061	10057	1.8733	-3482.4	-3097.8	98.82	29.10	57.54	723
110	36.0308	9552	1.8043	-3371.6	-2983.0	99.88	28.97	57.85	709
112	35.6500	9078	1.7352	-3259.8	-2867.1	100.93	28.87	58.10	693
114	35.2649	8537	1.6698	-3147.4	-2750.4	101.96	28.74	58.68	677
116	34.8697	8061	1.6056	-3033.4	-2631.9	102.98	28.55	59.06	662
118	34.4685	7595	1.5409	-2918.6	-2512.5	103.99	28.38	59.43	647
120	34.0592	7161	1.4806	-2802.6	-2391.6	105.00	28.22	59.89	632
122	33.6414	6756	1.4200	-2685.3	-2269.2	105.99	28.11	60.29	617
124	33.2171	6334	1.3610	-2567.0	-2145.6	106.98	28.05	60.92	602
126	32.7830	5938	1.3034	-2447.3	-2020.3	107.95	28.15	61.69	585
128	32.3394	5544	1.2469	-2325.2	-1892.3	108.93	28.01	62.33	570
130	31.8841	5174	1.1916	-2201.1	-1762.0	109.90	27.86	62.95	555
132	31.4172	4815	1.1380	-2074.5	-1628.8	110.87	27.48	63.44	541
134	30.9365	4475	1.0848	-1946.1	-1493.6	111.83	27.30	64.12	526
136	30.4463	4152	1.0327	-1495.1	-1035.2	112.79	27.07	64.75	511
138	29.9419	3831	0.9840	-1372.7	-905.1	113.74	26.77	65.67	497
140	29.4212	3532	0.9353	-1248.7	-772.9	114.69	26.63	66.69	482
142	28.8814	3247	0.8878	-1122.8	-638.0	115.64	26.59	67.91	467
144	28.3269	2974	0.8393	-995.5	-501.3	116.60	26.55	69.04	451
146	27.7530	2711	0.7938	-866.3	-361.8	117.56	26.57	70.63	436
148	27.1585	2472	0.7494	-734.7	-219.2	118.53	26.64	72.23	420
150	26.5415	2249	0.7050	-600.7	-73.2	119.51	26.63	73.70	405
152	25.9012	2035	0.6633	-464.4	76.1	120.50	26.61	75.60	390
154	25.2379	1840	0.6205	-325.8	228.9	121.50	26.65	77.24	375
156	24.5506	1663	0.5827	-184.5	385.7	122.51	26.66	79.51	361
158	23.8389	1503	0.5426	-40.8	546.5	123.54	26.60	81.08	347
160	23.1070	1361	0.5042	104.6	710.4	124.57	26.55	82.52	334
165	21.1933	1083	0.4209	471.0	1131.6	127.16	25.58	85.46	309
170	19.2435	913	0.3503	841.4	1568.6	129.77	25.78	87.51	286
175	17.3910	835	0.2935	1199.5	2004.5	132.30	25.75	85.45	270
180	15.7488	825	0.2498	1531.9	2420.6	134.64	25.44	80.36	262
185	14.3621	857	0.2163	1832.3	2807.1	136.76	25.10	74.05	258
190	13.2135	914	0.1907	2102.4	3161.9	138.65	24.77	68.07	257
195	12.2630	984	0.1706	2346.7	3488.3	140.35	24.49	62.84	258
200	11.4677	1060	0.1548	2570.5	3791.3	141.88	24.25	58.64	260
205	10.7954	1139	0.1416	2777.6	4074.5	143.28	24.06	55.01	262
210	10.2172	1218	0.1311	2972.1	4342.3	144.57	23.91	52.30	265
215	9.71597	1297	0.1218	3156.0	4596.9	145.77	23.80	49.83	267
220	9.27440	1375	0.1143	3331.6	4841.2	146.89	23.71	48.01	271
225	8.88330	1452	0.1074	3500.9	5076.5	147.95	23.65	46.30	273
230	8.53228	1527	0.1018	3663.8	5304.7	148.96	23.61	45.04	277
235	8.21601	1601	0.0964	3822.5	5526.5	149.91	23.59	43.82	280
240	7.92825	1673	0.0920	3977.3	5743.2	150.82	23.58	42.90	283
245	7.66541	1745	0.0879	4128.9	5955.3	151.70	23.58	42.04	286
250	7.42358	1815	0.0843	4277.8	6163.7	152.54	23.59	41.35	289
255	7.20085	1884	0.0808	4424.2	6368.4	153.35	23.60	40.63	292
260	6.99409	1951	0.0778	4568.6	6570.2	154.14	23.61	40.11	293
265	6.80179	2018	0.0750	4711.1	6769.4	154.89	23.63	39.59	298
270	6.62204	2083	0.0725	4852.0	6966.2	155.63	23.65	39.17	301
275	6.45388	2148	0.0700	4991.6	7160.8	156.34	23.67	38.73	304
280	6.29576	2212	0.0677	5129.8	7353.5	157.04	23.69	38.34	307
285	6.14698	2274	0.0657	5267.0	7544.5	157.72	23.71	38.04	310
290	6.00640	2336	0.0638	5403.1	7733.9	158.37	23.73	37.71	313
295	5.87295	2395	0.0623	5538.8	7922.6	159.02	23.77	37.64	316
300	5.74618	2457	0.0607	5673.9	8110.3	159.65	23.82	37.45	319

* TWO-PHASE BOUNDARY

TABLE 16. THERMODYNAMIC PROPERTIES OF FLUORINE

15 MN/m² ISOBAR

TEMPERATURE (IPTS 1968) K	DENSITY MOL/L	ISOTHERM DERIVATIVE J/MOL	ISOCHORE DERIVATIVE MN/m²-K	INTERNAL ENERGY J/MOL	ENTHALPY J/MOL	ENTROPY J/MOL-K	C_v J / MOL - K	C_p J / MOL - K	VELOCITY OF SOUND M/S
* 54.927	45.1635	30877	4.3510	-6338.8	-6006.6	61.18	36.11	52.62	1088
56	44.9980	30000	4.3482	-6283.3	-5950.0	62.23	35.92	53.35	1083
58	44.6915	28499	4.3430	-6179.6	-5844.0	64.12	35.56	54.77	1075
60	44.3817	28154	4.3044	-6075.0	-5737.0	65.98	35.26	55.31	1078
62	44.0781	27213	4.2139	-5970.8	-5630.5	67.76	35.36	56.18	1067
64	43.7773	25774	4.0430	-5865.8	-5523.2	69.50	35.53	56.71	1040
66	43.4660	24748	3.8472	-5759.7	-5414.6	71.21	35.36	56.25	1018
68	43.1517	24394	3.7358	-5652.9	-5305.3	72.88	34.71	55.60	1014
70	42.8449	23901	3.6392	-5547.9	-5197.8	74.47	33.97	55.10	1010
72	42.5428	22588	3.5558	-5444.0	-5091.5	76.00	33.54	55.81	995
74	42.2311	21725	3.4412	-5339.1	-4983.9	77.51	33.46	56.08	979
76	41.9141	20676	3.3150	-5232.9	-4875.0	78.99	33.35	56.34	959
78	41.5964	20018	3.1749	-5126.7	-4766.1	80.44	32.99	55.69	943
80	41.2778	19226	3.0698	-5021.4	-4658.0	81.84	32.64	55.65	929
82	40.9593	18422	2.9774	-4916.1	-4549.9	83.20	32.39	55.91	915
84	40.6302	17596	2.8634	-4809.4	-4440.3	84.55	32.06	55.77	898
86	40.3038	17005	2.7664	-4703.4	-4331.3	85.86	31.63	55.46	886
88	39.9761	16447	2.6797	-4597.3	-4222.1	87.15	31.14	55.18	876
90	39.6450	15646	2.5927	-4491.8	-4113.4	88.39	30.99	55.59	859
92	39.3122	15053	2.5121	-4386.1	-4004.5	89.61	30.96	55.92	846
94	38.9760	14388	2.4355	-4279.6	-3894.8	90.82	30.77	56.28	832
96	38.6334	13706	2.3505	-4172.5	-3784.2	92.00	30.33	56.26	818
98	38.2889	13084	2.2687	-4065.6	-3673.8	93.16	30.21	56.51	803
100	37.9403	12507	2.1876	-3957.9	-3562.6	94.30	30.10	56.68	787
102	37.5893	11938	2.1034	-3850.1	-3451.1	95.42	29.95	56.71	771
104	37.2346	11384	2.0328	-3742.1	-3339.2	96.52	29.70	56.93	758
106	36.8714	10761	1.9639	-3632.7	-3225.8	97.61	29.41	57.35	743
108	36.5044	10285	1.8911	-3522.7	-3111.8	98.68	29.11	57.29	730
110	36.1342	9783	1.8242	-3412.8	-2997.7	99.74	28.96	57.61	716
112	35.7588	9315	1.7555	-3302.0	-2882.5	100.78	28.86	57.84	701
114	35.3803	8794	1.6884	-3190.7	-2766.7	101.80	28.74	58.26	685
116	34.9919	8306	1.6253	-3077.9	-2649.2	102.82	28.56	58.69	670
118	34.5981	7842	1.5616	-2964.3	-2530.8	103.83	28.38	59.03	655
120	34.1966	7406	1.5010	-2849.6	-2411.0	104.82	28.22	59.43	641
122	33.7869	6998	1.4411	-2733.6	-2289.7	105.81	28.11	59.83	626
124	33.3719	6584	1.3821	-2616.8	-2167.3	106.79	28.05	60.35	611
126	32.9480	6189	1.3247	-2498.6	-2043.4	107.75	28.15	61.06	594
128	32.5158	5799	1.2689	-2378.1	-1916.8	108.72	28.01	61.63	579
130	32.0728	5427	1.2138	-2255.8	-1788.1	109.68	27.86	62.17	565
132	31.6196	5068	1.1611	-2131.2	-1656.8	110.64	27.47	62.59	551
134	31.1559	4728	1.1090	-2004.9	-1523.4	111.58	27.29	63.21	537
136	30.6801	4404	1.0575	-1527.0	-1038.1	112.53	27.05	63.75	523
138	30.1945	4088	1.0099	-1407.0	-910.3	113.46	26.69	64.46	510
140	29.6944	3789	0.9620	-1285.7	-780.6	114.39	26.58	65.36	495
142	29.1778	3501	0.9149	-1162.7	-648.6	115.33	26.55	66.42	480
144	28.6493	3230	0.8674	-1038.5	-515.0	116.26	26.53	67.39	465
146	28.1055	2966	0.8220	-913.0	-379.3	117.20	26.52	68.63	449
148	27.5434	2727	0.7792	-785.4	-240.8	118.14	26.58	70.02	435
150	26.9629	2501	0.7358	-655.8	-99.5	119.09	26.58	71.26	420
152	26.3646	2285	0.6944	-524.5	44.4	120.04	26.54	72.68	406
154	25.7477	2087	0.6532	-391.5	191.1	121.00	26.53	74.01	391
156	25.1116	1906	0.6162	-236.4	341.0	121.97	26.55	75.83	378
158	24.4566	1740	0.5773	-119.3	494.0	122.94	26.51	77.09	365
160	23.7856	1592	0.5398	19.0	649.6	123.92	26.40	78.16	352
165	22.0389	1289	0.4576	365.6	1046.2	126.36	25.41	80.58	328
170	20.2480	1086	0.3864	716.6	1457.4	128.82	25.69	82.73	303
175	18.5028	971	0.3273	1061.6	1872.3	131.22	25.70	82.13	286
180	16.8950	926	0.2803	1389.0	2276.9	133.50	25.44	78.96	275
185	15.4831	931	0.2433	1691.5	2660.3	135.60	25.11	74.16	269
190	14.2781	968	0.2144	1967.7	3018.2	137.51	24.80	69.04	266
195	13.2603	1024	0.1914	2219.7	3350.9	139.24	24.54	64.21	266
200	12.3991	1090	0.1732	2451.3	3661.1	140.81	24.32	60.11	266
205	11.6652	1163	0.1580	2666.1	3952.0	142.25	24.13	56.50	268
210	11.0322	1238	0.1459	2867.4	4227.0	143.58	23.99	53.68	270
215	10.4825	1314	0.1352	3057.4	4488.4	144.81	23.88	51.12	272
220	9.99820	1390	0.1266	3238.7	4738.9	145.96	23.79	49.19	275
225	9.56941	1465	0.1187	3412.3	4979.8	147.04	23.72	47.37	277
230	9.18492	1539	0.1123	3580.1	5213.2	148.07	23.68	46.02	281
235	8.83866	1612	0.1063	3742.7	5439.8	149.04	23.65	44.74	283
240	8.52394	1685	0.1013	3901.1	5660.9	149.97	23.64	43.75	286
245	8.23693	1756	0.0965	4055.9	5877.0	150.86	23.64	42.80	289
250	7.97303	1826	0.0925	4207.7	6089.0	151.72	23.64	42.07	292
255	7.73022	1895	0.0885	4356.8	6297.2	152.55	23.65	41.30	295
260	7.50510	1963	0.0852	4503.7	6502.3	133.34	23.67	40.73	298
265	7.29596	2030	0.0820	4648.5	6704.4	154.11	23.68	40.17	301
270	7.10063	2096	0.0792	4791.6	6904.1	154.86	23.70	39.73	304
275	6.91806	2161	0.0764	4933.2	7101.4	155.58	23.72	39.25	307
280	6.74656	2225	0.0739	5073.3	7296.7	156.29	23.74	38.83	310
285	6.58533	2289	0.0717	5212.2	7490.0	156.97	23.76	38.51	312
290	6.43311	2351	0.0695	5350.0	7681.7	157.64	23.77	38.15	315
295	6.28937	2413	0.0675	5486.7	7871.7	158.29	23.79	37.88	318
300	6.15331	2474	0.0657	5622.3	8060.0	158.92	23.81	37.63	321

* TWO-PHASE BOUNDARY

TABLE 16. THERMODYNAMIC PROPERTIES OF FLUORINE

16 MN/m² ISOBAR

TEMPERATURE (IPTS 1968) K	DENSITY MOL/L	ISOTHERM DERIVATIVE J/MOL	ISOCHORE DERIVATIVE MN/m²-K	INTERNAL ENERGY J/MOL	ENTHALPY J/MOL	ENTROPY J/MOL-K	C_v J / MOL - K	C_p	VELOCITY OF SOUND M/S
* 55.022	45.1813	31058	4.3518	-6359.3	-6005.2	61.21	36.10	52.53	1091
56	45.0312	30232	4.3492	-6309.0	-5953.7	62.15	35.92	53.20	1086
58	44.7265	28661	4.3439	-6205.9	-5848.1	64.05	35.55	54.64	1077
60	44.4171	28308	4.3152	-6101.6	-5741.4	65.90	35.26	55.26	1081
62	44.1147	27477	4.2249	-5997.8	-5635.2	67.68	35.36	56.06	1071
64	43.8158	26124	4.0639	-5893.3	-5528.1	69.42	35.56	56.64	1046
66	43.5062	25039	3.8508	-5787.4	-5419.7	71.13	35.43	56.08	1021
68	43.1926	24565	3.7362	-5681.0	-531u.6	72.79	34.79	55.50	1016
70	42.8865	24108	3.6378	-5576.3	-5203.2	74.39	34.04	54.93	1012
72	42.5868	22892	3.5604	-5473.0	-5097.3	75.91	33.58	55.57	998
74	42.2767	22082	3.4523	-5368.5	-4990.0	77.42	33.51	55.85	984
76	41.9621	20950	3.3289	-5262.8	-4881.5	78.90	33.41	56.24	963
78	41.6460	20276	3.1857	-5157.0	-4772.8	80.35	33.07	55.58	947
80	41.3295	19486	3.0727	-5052.1	-4665.0	81.75	32.73	55.42	932
82	41.0131	18768	2.9858	-4947.4	-4557.2	83.11	32.46	55.62	920
84	40.6867	17810	2.8725	-4841.2	-4448.0	84.46	32.15	55.66	901
86	40.3622	17231	2.7711	-4735.7	-4339.3	85.77	31.72	55.25	889
88	40.0365	16706	2.6857	-4630.1	-4230.4	87.04	31.22	54.93	879
90	39.7085	15845	2.6007	-4525.2	-4122.2	88.29	31.07	55.43	863
92	39.3781	15296	2.5198	-4420.0	-4013.7	89.51	31.03	55.66	850
94	39.0448	14652	2.4462	-4314.2	-3904.4	90.71	30.83	56.01	837
96	38.7058	13935	2.3361	-4207.8	-3794.4	91.89	30.37	56.11	823
98	38.3646	13319	2.2860	-4101.6	-3684.6	93.04	30.22	56.34	808
100	38.0195	12737	2.2072	-3994.6	-3573.8	94.18	30.09	56.55	794
102	37.6722	12184	2.1220	-3887.5	-3462.8	95.29	29.97	56.53	778
104	37.3214	11644	2.0460	-3780.2	-3351.5	96.39	29.72	56.61	764
106	36.9634	10984	1.9837	-3671.8	-3238.9	97.47	29.39	57.18	750
108	36.6006	10512	1.9107	-3562.6	-3125.5	98.55	29.10	57.10	737
110	36.2353	10011	1.8435	-3453.7	-3012.1	99.59	28.94	57.38	723
112	35.8648	9546	1.7765	-3343.7	-2897.6	100.63	28.84	57.62	709
114	35.4924	9055	1.7069	-3233.4	-2782.6	101.65	28.73	57.85	693
116	35.1106	8551	1.6446	-3121.8	-2666.0	102.66	28.55	58.32	678
118	34.7236	8091	1.5822	-3009.3	-2548.5	103.66	28.37	58.65	663
120	34.3294	7651	1.5211	-2895.8	-2429.7	104.65	28.21	59.00	649
122	33.9274	7238	1.4618	-2781.1	-2309.5	105.63	28.10	59.40	634
124	33.5210	6832	1.4029	-2665.6	-2188.2	106.60	28.04	59.83	619
126	33.1064	6437	1.3460	-2548.8	-2065.5	107.56	28.15	60.50	603
128	32.6846	6050	1.2907	-2429.9	-1940.4	108.51	28.01	61.00	589
130	32.2530	5676	1.2361	-2309.2	-1813.2	109.47	27.86	61.50	574
132	31.8122	5317	1.1832	-2186.4	-1683.4	110.41	27.47	61.81	561
134	31.3620	4976	1.1323	-2062.0	-1551.8	111.35	27.28	62.39	547
136	30.9010	4650	1.0814	-1557.3	-1039.5	112.28	27.04	62.86	533
138	30.4319	4339	1.0328	-1439.4	-913.6	113.20	26.65	63.27	521
140	29.9499	4041	0.9892	-1320.5	-786.3	114.11	26.50	64.31	508
142	29.4537	3750	0.9398	-1200.0	-656.8	115.03	26.51	65.07	492
144	28.9474	3481	0.8948	-1078.6	-525.8	115.95	26.53	66.06	478
146	28.4292	3215	0.8482	-956.1	-393.3	116.86	26.50	66.93	462
148	27.8943	2976	0.8063	-831.9	-258.3	117.78	26.53	68.08	448
150	27.3443	2746	0.7649	-706.0	-120.9	118.70	26.56	69.31	434
152	26.7803	2529	0.7233	-578.8	18.7	119.62	26.50	70.34	420
154	26.2009	2329	0.6833	-450.2	160.4	120.55	26.46	71.43	407
156	25.6058	2145	0.6471	-320.1	304.7	121.48	26.47	72.92	394
158	24.9955	1975	0.6091	-188.4	451.7	122.42	26.45	73.97	381
160	24.3725	1821	0.5724	-55.7	600.7	123.36	26.36	74.84	369
165	22.7580	1498	0.4913	275.3	978.3	125.68	25.25	76.58	346
170	21.0999	1268	0.4200	610.5	1368.8	128.01	25.61	78.74	320
175	19.4611	1122	0.3596	942.7	1764.9	130.31	25.65	78.87	301
180	17.9117	1047	0.3101	1262.9	2156.2	132.51	25.45	76.99	289
185	16.5081	1025	0.2702	1563.5	2532.7	134.58	25.11	73.47	281
190	15.2760	1040	0.2382	1842.2	2889.6	136.48	24.82	69.27	276
195	14.2123	1080	0.2127	2099.3	3225.1	138.22	24.56	65.01	274
200	13.2996	1134	0.1919	2336.9	3540.0	139.82	24.35	61.10	274
205	12.5132	1198	0.1751	2558.0	3636.7	141.28	24.18	57.67	274
210	11.8312	1267	0.1613	2765.4	4117.7	142.64	24.04	54.84	276
215	11.2368	1339	0.1492	2961.0	4384.9	143.90	23.93	52.24	277
220	10.7124	1412	0.1394	3147.3	4640.9	145.07	23.85	5C.24	280
225	10.2477	1485	0.1305	3325.7	4887.0	146.18	23.79	48.35	282
230	9.83097	1558	0.1232	3497.6	5125.1	147.23	23.74	46.93	285
235	9.45564	1630	0.1165	3664.0	5356.1	148.22	23.71	45.59	287
240	9.11462	1702	0.1108	3825.9	5581.3	149.17	23.70	44.54	290
245	8.80411	1772	0.1054	3983.8	5801.1	150.08	23.69	43.50	293
250	8.51863	1841	0.1009	4138.5	6016.7	150.95	23.69	42.74	296
255	8.25593	1910	0.0965	4290.2	6228.2	151.78	23.70	41.94	298
260	8.01259	1979	0.0928	4439.5	6436.3	152.59	23.71	41.33	301
265	7.78678	2046	0.0892	4586.5	6643.3	153.37	23.73	40.73	304
270	7.57601	2112	0.0861	4731.8	6843.7	154.13	23.75	40.25	307
275	7.37912	2177	0.0830	4875.3	7043.6	154.86	23.77	39.75	310
280	7.19429	2242	0.0802	5017.3	7241.3	155.58	23.78	39.30	312
285	7.02069	2306	0.0777	5157.9	7436.9	156.27	23.80	38.95	315
290	6.85688	2369	0.0753	5297.3	7630.8	156.94	23.82	38.60	318
295	6.70223	2432	0.0732	5435.6	7822.9	157.60	23.84	38.29	321
300	6.55600	2493	0.0711	5572.7	8013.2	158.24	23.85	38.02	323

* TWO-PHASE BOUNDARY

TABLE 16. THERMODYNAMIC PROPERTIES OF FLUORINE

17 MN/m² ISOBAR

TEMPERATURE (IPTS 1968) K	DENSITY MOL/L	ISOTHERM DERIVATIVE J/MOL	ISOCHORE DERIVATIVE MN/m²-K	INTERNAL ENERGY J/MOL	ENTHALPY J/MOL	ENTROPY J/MOL-K	C_v	C_p J / MOL - K	VELOCITY OF SOUND M/S
* 55.116	45.1991	31232	4.3523	-6379.9	-6003.8	61.23	36.08	52.44	1093
56	45.0642	30462	4.3500	-6334.6	-5957.4	62.08	35.92	53.05	1088
58	44.7613	28822	4.3446	-6232.0	-5852.2	63.97	35.55	54.51	1078
60	44.4523	28461	4.3258	-6128.1	-5745.7	65.82	35.26	55.22	1083
62	44.1509	27756	4.2357	-6024.8	-5639.7	67.60	35.36	55.92	1075
64	43.8538	26503	4.0864	-5920.6	-5533.0	69.34	35.59	56.56	1053
66	43.5459	25353	3.8666	-5815.0	-5424.6	71.05	35.49	56.02	1026
68	43.2331	24743	3.7370	-5709.0	-5315.7	72.71	34.87	55.40	1017
70	42.9278	24321	3.6350	-5604.5	-5208.5	74.31	34.11	54.75	1014
72	42.6302	23210	3.5633	-5501.8	-5103.0	75.83	33.63	55.30	1002
74	42.3216	22462	3.4632	-5397.7	-4996.0	77.33	33.55	55.61	990
76	42.0095	21240	3.3429	-5292.6	-4887.9	78.81	33.45	56.11	968
78	41.6950	20545	3.1975	-5187.1	-4779.4	80.26	33.15	55.48	951
80	41.3805	19758	3.0749	-5082.6	-4671.8	81.65	32.81	55.17	935
82	41.0659	19132	2.9907	-4978.4	-4564.4	83.01	32.53	55.26	925
84	40.7425	18031	2.8819	-4872.8	-4455.6	84.36	32.23	55.54	904
86	40.4199	17463	2.7753	-4767.7	-4347.1	85.67	31.81	55.03	892
88	40.0959	16970	2.6913	-4662.6	-4238.6	86.95	31.31	54.67	883
90	39.7712	16043	2.6081	-4558.3	-4130.9	88.18	31.15	55.27	866
92	39.4430	15542	2.5265	-4453.7	-4022.7	89.40	31.10	55.39	854
94	39.1125	14921	2.4555	-4348.5	-3913.9	90.60	30.88	55.71	842
96	38.7770	14165	2.3803	-4242.8	-3804.4	91.77	30.41	55.94	828
98	38.4391	13555	2.3032	-4137.3	-3695.1	92.92	30.22	56.17	814
100	38.0973	12968	2.2261	-4031.0	-3584.8	94.06	30.08	56.41	800
102	37.7535	12432	2.1418	-3924.6	-3474.3	95.17	29.97	56.38	785
104	37.4063	11907	2.0619	-3817.9	-3363.5	96.27	29.73	56.27	770
106	37.0535	11205	2.0024	-3710.5	-3251.7	97.34	29.36	56.98	757
108	36.6947	10736	1.9310	-3602.2	-3138.9	98.41	29.08	56.94	744
110	36.3340	10237	1.8623	-3494.0	-3026.1	99.45	28.92	57.15	730
112	35.9683	9771	1.7983	-3385.6	-2912.4	100.49	28.80	57.45	716
114	35.6012	9320	1.7281	-3275.6	-2798.1	101.50	28.72	57.54	701
116	35.2259	8796	1.6639	-3165.0	-2682.4	102.51	28.55	57.97	686
118	34.8453	8342	1.6023	-3053.6	-2565.8	103.50	28.36	58.27	672
120	34.4580	7897	1.5410	-2941.3	-2447.9	104.48	28.19	58.58	657
122	34.0633	7474	1.4823	-2827.7	-2328.7	105.46	28.09	59.00	643
124	33.6648	7078	1.4242	-2713.4	-2208.5	106.42	28.03	59.39	628
126	33.2589	6683	1.3671	-2598.1	-2086.9	107.37	28.14	60.00	612
128	32.8465	6300	1.3118	-2480.6	-1963.0	108.32	28.00	60.41	598
130	32.4254	5924	1.2578	-2361.4	-1837.2	109.26	27.86	60.88	584
132	31.9960	5564	1.2039	-2240.2	-1708.9	110.20	27.47	61.06	570
134	31.5582	5221	1.1546	-2117.6	-1578.9	111.12	27.28	61.63	557
136	31.1107	4891	1.1046	-1586.1	-1039.6	112.04	27.03	62.08	544
138	30.6560	4587	1.0558	-1470.0	-915.5	112.94	26.63	62.32	532
140	30.1901	4288	1.0131	-1353.4	-790.3	113.85	26.42	63.18	520
142	29.7121	3993	0.9668	-1235.1	-662.9	114.75	26.47	64.12	505
144	29.2250	3726	0.9196	-1116.0	-534.3	115.65	26.54	64.80	489
146	28.7290	3460	0.8745	-996.2	-404.4	116.54	26.52	65.62	475
148	28.2172	3221	0.8316	-874.9	-272.4	117.44	26.52	66.42	461
150	27.6934	2985	0.7914	-752.2	-138.3	118.34	26.56	67.59	447
152	27.1581	2768	0.7509	-628.4	-2.4	119.24	26.50	68.49	434
154	26.6098	2565	0.7117	-503.6	135.3	120.14	26.41	69.35	421
156	26.0484	2378	0.6759	-377.5	275.1	121.04	26.40	70.55	409
158	25.4744	2205	0.6387	-250.2	417.1	121.95	26.40	71.45	396
160	24.8902	2047	0.6028	-122.1	560.9	122.85	26.34	72.18	384
165	23.3829	1707	0.5225	196.2	923.2	125.08	25.14	73.38	362
170	21.8354	1456	0.4514	518.4	1296.9	127.31	25.53	75.42	336
175	20.2937	1285	0.3900	839.3	1677.0	129.52	25.63	75.94	317
180	18.8112	1182	0.3388	1151.2	2054.9	131.65	25.43	74.82	303
185	17.4364	1134	0.2966	1448.2	2423.1	133.67	25.07	72.26	293
190	16.2005	1127	0.2621	1726.6	2776.0	135.55	24.82	68.92	287
195	15.1107	1150	0.2341	1986.4	3111.4	137.29	24.57	65.30	284
200	14.1809	1191	0.2111	2228.1	3428.6	138.90	24.38	61.68	282
205	13.3327	1245	0.1925	2454.1	3729.2	140.38	24.21	58.53	281
210	12.6089	1307	0.1770	2666.5	4014.8	141.76	24.08	55.76	282
215	11.9746	1373	0.1636	2867.1	4286.8	143.04	23.97	53.18	283
220	11.4134	1443	0.1526	3058.1	4547.5	144.24	23.89	51.15	285
225	10.9158	1513	0.1426	3240.8	4798.2	145.36	23.84	49.23	287
230	10.4678	1584	0.1344	3416.6	5040.7	146.43	23.79	47.75	289
235	10.0648	1654	0.1269	3586.6	5275.7	147.44	23.76	46.36	291
240	9.69849	1725	0.1206	3751.8	5504.7	148.41	23.75	45.27	294
245	9.36532	1793	0.1144	3912.7	5727.9	149.33	23.74	44.14	296
250	9.05902	1861	0.1095	4070.1	5946.7	150.21	23.74	43.35	299
255	8.77685	1930	0.1047	4224.4	6161.3	151.06	23.75	42.55	302
260	8.51561	1998	0.1006	4376.1	6372.4	151.88	23.76	41.90	305
265	8.27341	2065	0.0965	4525.3	6580.1	152.67	23.77	41.25	307
270	8.04746	2131	0.0931	4672.6	6785.0	153.44	23.79	40.75	310
275	7.83643	2197	0.0897	4818.0	6987.4	154.18	23.81	40.23	313
280	7.63843	2262	0.0866	4961.9	7187.4	154.90	23.83	39.75	315
285	7.45260	2325	0.0839	5104.2	7385.3	155.60	23.85	39.38	318
290	7.27732	2389	0.0814	5245.2	7581.2	156.28	23.87	39.04	321
295	7.11175	2453	0.0790	5385.1	7775.5	156.95	23.88	38.71	323
300	6.95534	2515	0.0767	5523.6	7967.7	157.60	23.90	38.41	326

* TWO-PHASE BOUNDARY

TABLE 16. THERMODYNAMIC PROPERTIES OF FLUORINE

ISOTHERM DERIVATIVE J/MOL	ISOCHORE DERIVATIVE MN/m²-K	INTERNAL ENERGY J/MOL	ENTHALPY J/MOL	ENTROPY J/MOL-K	C_v J / MOL - K	C_p	VELOCITY OF SOUND M/S
31399	4.3527	-6400.4	-6002.3	61.25	36.06	52.36	1095
30690	4.3505	-6360.2	-5961.0	62.01	35.92	52.90	1091
28982	4.3452	-6258.1	-5856.3	63.90	35.55	54.38	1080
28613	4.3361	-6154.6	-5750.0	65.75	35.26	55.18	1086
28051	4.2463	-6051.6	-5644.2	67.52	35.36	55.77	1079
26910	4.1102	-5947.9	-5537.8	69.26	35.61	56.47	1060
25689	3.8842	-5842.6	-5429.6	70.97	35.55	55.95	1032
24928	3.7381	-5736.8	-5320.8	72.63	34.94	55.30	1019
24539	3.6307	-5632.6	-5213.7	74.23	34.19	54.55	1015
23543	3.5646	-5530.4	-5108.6	75.75	33.68	55.02	1006
22864	3.4740	-5426.8	-5001.9	77.25	33.59	55.35	996
21547	3.3542	-5322.2	-4894.2	78.72	33.50	55.93	973
20826	3.2102	-5217.0	-4785.8	80.17	33.22	55.37	956
20039	3.0763	-5112.9	-4678.5	81.56	32.89	54.90	938
19512	2.9946	-5009.1	-4571.4	82.92	32.60	54.90	930
18258	2.8916	-4904.2	-4463.0	84.26	32.31	55.42	908
17899	2.7791	-4799.5	-4354.8	85.57	31.90	54.81	895
17241	2.6964	-4694.8	-4246.5	86.85	31.40	54.41	887
16240	2.6149	-4591.3	-4139.4	88.08	31.23	55.11	868
15791	2.5327	-4487.1	-4031.5	89.30	31.17	55.12	857
15195	2.4637	-4382.6	-3923.1	90.49	30.94	55.40	846
14397	2.3929	-4277.5	-3814.2	91.66	30.45	55.75	833
13794	2.3202	-4172.8	-3705.4	92.81	30.21	56.00	820
13199	2.2450	-4067.2	-3595.7	93.94	30.06	56.26	806
12681	2.1629	-3961.4	-3485.6	95.05	29.96	56.25	792
12171	2.0744	-3855.3	-3375.2	96.14	29.75	55.91	776
11426	2.0190	-3748.9	-3264.3	97.21	29.33	56.75	763
10959	1.9520	-3641.3	-3152.0	98.28	29.05	56.80	751
10461	1.8805	-3534.0	-3039.9	99.32	28.90	56.92	736
9991	1.8193	-3425.8	-2926.8	100.35	28.75	57.27	724
9590	1.7502	-3317.2	-2813.1	101.36	28.69	57.25	710
9041	1.6831	-3207.7	-2698.3	102.36	28.53	57.63	693
8595	1.6216	-3097.3	-2582.5	103.34	28.35	57.88	680
8143	1.5615	-2986.1	-2465.6	104.32	28.17	58.21	666
7707	1.5029	-2873.7	-2347.3	105.29	28.07	58.65	651
7324	1.4451	-2760.5	-2228.0	106.24	28.02	58.96	637
6927	1.3879	-2646.4	-2107.6	107.19	28.13	59.53	621
6547	1.3320	-2530.2	-1984.8	108.13	28.00	59.85	607
6168	1.2791	-2412.5	-1860.2	109.06	27.65	60.32	593
5807	1.2259	-2292.8	-1733.3	109.99	27.47	60.48	580
5462	1.1757	-2171.8	-1604.8	110.90	27.27	60.93	567
5128	1.1271	-1613.6	-1038.7	111.81	27.02	61.39	554
4831	1.0787	-1499.2	-916.0	112.71	26.62	61.50	542
4531	1.0332	-1384.4	-792.6	113.59	26.37	62.02	530
4231	0.9949	-1268.4	-667.5	114.48	26.38	63.40	517
3967	0.9440	-1151.2	-540.8	115.37	26.52	63.73	501
3701	0.8981	-1033.7	-413.2	116.25	26.59	64.41	486
3462	0.8550	-914.8	-283.5	117.13	26.52	64.95	472
3219	0.8157	-795.0	-152.5	118.01	26.57	66.07	459
3001	0.7763	-674.1	-19.7	118.89	26.55	66.90	446
2797	0.7374	-552.4	114.7	119.77	26.41	67.53	434
2608	0.7029	-429.9	250.6	120.64	26.33	68.58	423
2432	0.6665	-306.3	388.6	121.52	26.35	69.35	410
2270	0.6312	-181.9	528.1	122.40	26.32	70.00	399
1916	0.5515	125.9	877.9	124.55	25.07	70.78	377
1647	0.4807	437.0	1237.7	126.70	25.47	72.65	352
1454	0.4188	748.0	1604.1	128.82	25.59	73.35	331
1328	0.3663	1052.0	1969.9	130.89	25.37	72.66	316
1257	0.3223	1344.1	2329.1	132.85	25.00	70.79	306
1229	0.2857	1620.6	2676.3	134.71	24.68	68.06	299
1233	0.2555	1881.2	3009.7	136.44	24.57	65.15	293
1263	0.2306	2125.6	3327.4	138.05	24.39	61.93	291
1304	0.2102	2355.0	3630.0	139.54	24.24	59.09	289
1356	0.1931	2571.4	3918.7	140.93	24.12	56.46	289
1417	0.1783	2776.2	4194.5	142.23	24.02	53.96	289
1481	0.1660	2971.3	4459.1	143.45	23.93	51.92	291
1548	0.1551	3157.7	4713.7	144.59	23.87	49.99	292
1616	0.1460	3337.4	4960.0	145.68	23.84	48.48	294
1684	0.1377	3510.8	5198.7	146.70	23.81	47.06	296
1753	0.1306	3679.1	5431.1	147.68	23.79	45.93	298
1821	0.1238	3842.8	5657.5	148.62	23.78	44.73	300
1888	0.1182	4002.6	5879.3	149.51	23.79	43.90	303
1955	0.1131	4159.6	6096.8	150.37	23.79	43.12	305
2022	0.1085	4313.6	6310.7	151.20	23.80	42.44	308
2089	0.1041	4465.0	6520.9	152.01	23.82	41.75	310
2155	0.1003	4614.2	6728.3	152.78	23.83	41.21	313
2220	0.0966	4761.4	6932.9	153.53	23.85	40.68	316
2284	0.0932	4907.0	7135.2	154.26	23.87	40.17	318
2348	0.0902	5051.0	7335.1	154.97	23.89	39.79	321
2411	0.0875	5193.7	7533.1	155.66	23.91	39.47	324
2475	0.0849	5335.0	7729.4	156.33	23.93	39.13	326
2539	0.0824	5475.0	7923.6	156.98	23.94	38.81	329

TABLE 16. THERMODYNAMIC PROPERTIES OF FLUORINE

19 MN/m² ISOBAR

TEMPERATURE (IPTS 1968) K	DENSITY MOL/L	ISOTHERM DERIVATIVE J/MOL	ISOCHORE DERIVATIVE MN/m²-K	INTERNAL ENERGY J/MOL	ENTHALPY J/MOL	ENTROPY J/MOL-K	C_v J/MOL-K	C_p J/MOL-K	VELOCITY OF SOUND M/S
* 55.305	45.2344	31560	4.3528	-6420.9	-6000.9	61.28	36.05	52.27	1097
56	45.1294	30917	4.3509	-6385.7	-5964.6	61.94	35.92	52.76	1093
58	44.8303	29142	4.3455	-6284.1	-5860.3	63.82	35.55	54.25	1082
60	44.5222	28764	4.3400	-6181.0	-5754.3	65.67	35.26	55.08	1087
62	44.2222	28361	4.2565	-6078.3	-5648.7	67.45	35.36	55.62	1083
64	43.9281	27343	4.1353	-5975.0	-5542.5	69.18	35.63	56.37	1067
66	43.6237	26046	3.9032	-5870.0	-5434.5	70.89	35.60	55.89	1037
68	43.3134	25120	3.7395	-5764.6	-5325.9	72.55	35.02	55.20	1021
70	43.0093	24762	3.6266	-5660.7	-5218.9	74.15	34.27	54.37	1017
72	42.7151	23890	3.5643	-5558.8	-5114.0	75.66	33.73	54.72	1010
74	42.4091	23287	3.4845	-5455.7	-5007.7	77.16	33.63	55.08	1002
76	42.1023	21870	3.3654	-5351.6	-4900.3	78.64	33.54	55.75	978
78	41.7910	21117	3.2237	-5246.9	-4792.2	80.08	33.29	55.27	961
80	41.4803	20331	3.0769	-5143.1	-4685.0	81.48	32.98	54.63	941
82	41.1684	19908	2.9976	-5039.7	-4578.1	82.83	32.68	54.52	935
84	40.8521	18491	2.9014	-4935.5	-4470.4	84.17	32.39	55.30	912
86	40.5329	17941	2.7861	-4831.1	-4362.3	85.48	31.99	54.64	898
88	40.2119	17516	2.7011	-4726.8	-4254.3	86.75	31.48	54.15	890
90	39.8944	16436	2.6211	-4624.0	-4147.7	87.98	31.31	54.95	871
92	39.5696	16042	2.5405	-4520.3	-4040.1	89.20	31.25	54.89	861
94	39.2441	15473	2.4710	-4416.3	-3932.2	90.39	31.00	55.09	851
96	38.9159	14629	2.4039	-4312.0	-3823.8	91.55	30.50	55.53	837
98	38.5841	14033	2.3360	-4208.0	-3715.5	92.70	30.21	55.80	826
100	38.2488	13430	2.2637	-4103.0	-3606.3	93.83	30.04	56.12	813
102	37.9112	12931	2.1853	-3997.9	-3496.7	94.93	29.93	56.14	799
104	37.5707	12437	2.0939	-3892.3	-3386.6	96.02	29.76	55.73	783
106	37.2285	11646	2.0342	-3786.9	-3276.6	97.09	29.31	56.49	769
108	36.8772	11181	1.9739	-3680.2	-3164.9	98.15	29.01	56.68	758
110	36.5253	10683	1.8992	-3573.5	-3053.3	99.18	28.87	56.71	743
112	36.1686	10204	1.8390	-3466.3	-2940.9	100.21	28.71	57.08	731
114	35.8098	9865	1.7730	-3358.4	-2827.8	101.21	28.65	56.98	719
116	35.4472	9288	1.7023	-3249.8	-2713.8	102.21	28.51	57.31	701
118	35.0781	8851	1.6410	-3140.4	-2598.8	103.19	28.34	57.51	688
120	34.7037	8389	1.5818	-3030.2	-2482.7	104.16	28.14	57.86	674
122	34.3229	7937	1.5233	-2918.9	-2365.4	105.12	28.05	58.32	659
124	33.9380	7569	1.4656	-2806.8	-2247.0	106.07	28.00	58.55	645
126	33.5477	7169	1.4086	-2693.9	-2127.5	107.01	28.11	59.10	630
128	33.1522	6793	1.3532	-2578.9	-2005.8	107.94	27.99	59.39	616
130	32.7498	6411	1.2999	-2462.6	-1882.4	108.87	27.85	59.79	602
132	32.3406	6048	1.2475	-2344.3	-1756.8	109.79	27.47	59.94	589
134	31.9246	5699	1.1957	-2224.8	-1629.6	110.69	27.27	60.25	576
136	31.5010	5361	1.1487	-1639.9	-1036.8	111.59	27.01	60.75	563
138	31.0704	5072	1.1007	-1526.9	-915.4	112.48	26.61	60.75	552
140	30.6320	4771	1.0542	-1413.8	-793.5	113.35	26.35	61.11	540
142	30.1854	4465	1.0166	-1300.0	-670.6	114.22	26.29	62.36	528
144	29.7299	4203	0.9721	-1184.6	-545.5	115.10	26.48	63.10	513
146	29.2702	3939	0.9223	-1069.0	-419.8	115.97	26.65	63.45	497
148	28.7959	3700	0.8776	-952.0	-292.2	116.83	26.59	63.75	483
150	28.3160	3448	0.8388	-834.9	-163.9	117.70	26.58	64.76	470
152	27.8261	3229	0.8000	-716.6	-33.8	118.56	26.61	65.52	457
154	27.3266	3024	0.7617	-597.6	97.7	119.42	26.48	66.04	446
156	26.8174	2833	0.7263	-476.0	230.5	120.27	26.30	66.69	435
158	26.2994	2655	0.6927	-357.6	364.8	121.13	26.28	67.56	424
160	25.7742	2490	0.6579	-236.4	500.7	121.98	26.31	68.17	412
165	24.4308	2124	0.5786	62.6	840.3	124.07	25.02	68.59	391
170	23.0549	1840	0.5082	364.2	1188.3	126.15	25.45	70.35	366
175	21.6748	1628	0.4461	666.4	1543.0	128.21	25.52	71.05	345
180	20.3219	1482	0.3927	963.1	1898.0	130.21	25.20	70.55	330
185	19.0312	1390	0.3472	1249.7	2248.1	132.13	24.89	69.20	319
190	17.8290	1343	0.3088	1523.2	2588.9	133.94	24.45	66.91	311
195	16.7329	1329	0.2768	1783.0	2918.5	135.66	24.37	64.51	304
200	15.7433	1350	0.2508	2029.5	3236.4	137.27	24.35	61.97	301
205	14.8654	1375	0.2280	2261.1	3539.3	138.76	24.25	59.33	298
210	14.0822	1417	0.2094	2480.5	3829.7	140.16	24.16	56.92	296
215	13.3855	1469	0.1932	2688.7	4108.2	141.47	24.07	54.56	296
220	12.7632	1527	0.1798	2887.2	4375.9	142.70	23.98	52.57	297
225	12.2063	1590	0.1678	3077.2	4633.7	143.86	23.92	50.67	298
230	11.7048	1655	0.1578	3259.8	4883.1	144.96	23.86	49.11	299
235	11.2516	1721	0.1486	3436.5	5125.2	146.00	23.85	47.68	301
240	10.8390	1787	0.1409	3607.9	5360.6	146.99	23.84	46.53	303
245	10.4627	1857	0.1336	3774.4	5590.3	147.94	23.83	45.33	305
250	10.1178	1922	0.1273	3936.8	5814.7	148.85	23.83	44.40	307
255	9.79938	1986	0.1217	4095.9	6034.8	149.72	23.83	43.63	309
260	9.50426	2052	0.1167	4252.1	6251.2	150.56	23.85	42.94	312
265	9.23036	2119	0.1119	4405.6	6464.0	151.37	23.86	42.24	314
270	8.97529	2184	0.1077	4556.7	6673.6	152.15	23.87	41.66	317
275	8.73711	2248	0.1036	4705.7	6880.3	152.91	23.89	41.11	319
280	8.51369	2312	0.0999	4852.9	7084.6	153.65	23.91	40.59	321
285	8.30405	2375	0.0966	4998.5	7286.5	154.36	23.93	40.17	324
290	8.10649	2438	0.0937	5142.6	7486.4	155.06	23.95	39.84	327
295	7.91959	2500	0.0909	5285.5	7684.6	155.74	23.97	39.51	329
300	7.74302	2564	0.0883	5426.9	7880.7	156.40	23.99	39.19	332

* TWO-PHASE BOUNDARY

TABLE 16. THERMODYNAMIC PROPERTIES OF FLUORINE

4N/m² ISOBAR

DENSITY	ISOTHERM DERIVATIVE	ISOCHORE DERIVATIVE	INTERNAL ENERGY	ENTHALPY	ENTROPY	C_v	C_p	VELOCITY OF SOUND
MOL/L	J/MOL	MN/m²-K	J/MOL	J/MOL	J/MOL-K	J / MOL - K		M/S
45.2520	31714	4.3528	-6441.4	-5999.4	61.30	36.03	52.19	1100
45.1616	31142	4.3511	-6411.1	-5968.2	61.88	35.92	52.61	1096
44.8645	29300	4.3457	-6310.0	-5864.2	63.75	35.55	54.13	1083
44.5569	28915	4.3401	-6207.3	-5758.5	65.59	35.26	54.95	1089
44.2573	28685	4.2665	-6104.9	-5653.0	67.37	35.37	55.45	1088
43.9644	27800	4.1616	-6002.0	-5547.1	69.10	35.64	56.26	1075
43.6619	26421	3.9238	-5897.3	-5439.3	70.81	35.65	55.83	1043
43.3530	25318	3.7413	-5792.2	-5330.9	72.47	35.09	55.09	1023
43.0495	24990	3.6273	-5688.6	-5224.0	74.07	34.34	54.23	1019
42.7567	24249	3.5625	-5587.1	-5119.4	75.58	33.79	54.40	1014
42.4516	23728	3.4949	-5484.4	-5013.3	77.08	33.66	54.80	1008
42.1477	22207	3.3764	-5380.9	-4906.4	78.55	33.59	55.55	983
41.8380	21417	3.2381	-5276.5	-4798.5	79.99	33.35	55.17	966
41.5291	20633	3.0840	-5173.1	-4691.5	81.39	33.07	54.45	946
41.2181	20317	2.9998	-5070.0	-4584.8	82.75	32.75	54.13	940
40.9058	18730	2.9115	-4966.5	-4477.6	84.08	32.46	55.18	915
40.5882	18186	2.7956	-4862.5	-4369.8	85.38	32.08	54.51	902
40.2685	17796	2.7053	-4758.6	-4262.0	86.66	31.56	53.88	894
39.9548	16632	2.6268	-4656.5	-4155.9	87.88	31.39	54.78	874
39.6315	16296	2.5477	-4553.2	-4048.6	89.10	31.32	54.65	865
39.3082	15754	2.4773	-4449.8	-3941.0	90.28	31.06	54.76	855
38.9837	14863	2.4129	-4346.3	-3833.3	91.45	30.54	55.29	841
38.6547	14274	2.3500	-4242.8	-3725.4	92.59	30.21	55.59	831
38.3227	13661	2.2821	-4138.6	-3616.8	93.71	30.01	55.97	819
37.9878	13183	2.2091	-4034.1	-3507.6	94.82	29.90	56.06	807
37.6502	12704	2.1156	-3929.1	-3397.8	95.90	29.75	55.60	790
37.3136	11864	2.0479	-3824.6	-3288.6	96.96	29.30	56.21	774
36.9658	11401	1.9966	-3718.6	-3177.6	98.02	28.96	56.59	766
36.6179	10903	1.9214	-3612.7	-3066.6	99.05	28.84	56.62	751
36.2656	10410	1.8580	-3506.3	-2954.8	100.07	28.66	56.90	738
35.9098	10145	1.7967	-3399.1	-2842.1	101.08	28.59	56.72	728
35.5534	9534	1.7236	-3291.5	-2728.9	102.06	28.48	57.08	709
35.1894	9109	1.6606	-3182.9	-2614.6	103.04	28.32	57.17	696
34.8212	8635	1.6013	-3073.7	-2499.4	104.01	28.12	57.51	682
34.4471	8165	1.5436	-2963.6	-2383.0	104.96	28.02	58.02	667
34.0681	7812	1.4860	-2852.5	-2265.4	105.91	27.97	58.18	654
33.6849	7408	1.4296	-2740.6	-2146.9	106.84	28.09	58.73	638
33.2968	7036	1.3741	-2626.8	-2026.1	107.77	27.97	58.95	625
32.9029	6651	1.3199	-2511.7	-1903.9	108.68	27.84	59.29	611
32.5028	6286	1.2685	-2394.7	-1779.4	109.59	27.46	59.45	598
32.0966	5934	1.2163	-2276.6	-1653.5	110.49	27.27	59.78	585
31.6837	5589	1.1691	-2157.2	-1665.2	111.38	27.00	60.14	572
31.2630	5309	1.1225	-1553.5	-913.7	112.26	26.68	60.11	562
30.8366	5007	1.0761	-1441.9	-793.3	113.12	26.34	60.39	550
30.4037	4695	1.0341	-1330.0	-672.2	113.98	26.24	61.22	537
29.9615	4436	1.0010	-1216.5	-549.0	114.84	26.39	62.62	526
29.5168	4175	0.9487	-1102.4	-424.8	115.70	26.65	62.77	509
29.0579	3935	0.8967	-987.0	-298.7	116.56	26.75	62.57	492
28.5970	3671	0.8602	-872.4	-173.0	117.40	26.61	63.57	480
28.1255	3453	0.8228	-756.3	-45.2	118.25	26.66	64.33	468
27.6456	3247	0.7835	-639.5	83.9	119.09	26.61	64.69	456
27.1572	3055	0.7479	-522.5	213.9	119.93	26.36	65.08	446
26.6611	2875	0.7172	-405.0	345.1	120.77	26.20	65.97	436
26.1592	2708	0.6632	-286.7	477.8	121.60	26.25	66.56	425
24.8801	2330	0.6041	5.0	808.8	123.64	24.98	66.73	405
23.5718	2033	0.5342	298.4	1146.9	125.66	25.45	68.41	379
22.2581	1804	0.4718	592.7	1491.3	127.65	25.48	69.06	359
20.9627	1642	0.4180	882.7	1836.8	129.60	25.07	68.67	344
19.7170	1530	0.3712	1163.8	2178.1	131.47	24.66	67.51	332
18.5419	1466	0.3316	1434.0	2512.6	133.25	24.27	65.72	323
17.4571	1436	0.2977	1691.1	2836.7	134.94	23.95	63.45	316
16.4585	1450	0.2719	1939.2	3154.4	136.55	24.17	61.81	312
15.5720	1458	0.2459	2172.4	3456.7	138.04	24.20	59.24	307
14.7716	1487	0.2257	2394.0	3747.9	139.44	24.17	57.15	304
14.0530	1529	0.2083	2604.8	4028.0	140.76	24.11	54.98	303
13.4071	1581	0.1937	2806.3	4298.0	142.00	24.05	53.08	303
12.8261	1638	0.1807	2999.2	4558.5	143.17	23.98	51.25	304
12.3013	1700	0.1698	3184.8	4810.7	144.28	23.93	49.70	305
11.8259	1763	0.1598	3363.9	5055.1	145.33	23.88	48.22	306
11.3927	1827	0.1513	3538.0	5293.5	146.34	23.87	47.05	308
10.9949	1903	0.1441	3707.5	5526.6	147.30	23.88	46.01	311
10.6322	1968	0.1368	3872.4	5753.5	148.22	23.88	44.91	312
10.2980	2026	0.1305	4033.5	5975.6	149.10	23.87	44.07	314
9.98757	2088	0.1250	4191.6	6194.1	149.95	23.88	43.38	316
9.69830	2157	0.1201	4347.2	6409.4	150.77	23.90	42.75	319
9.42947	2221	0.1153	4500.2	6621.2	151.56	23.92	42.10	321
9.17867	2283	0.1109	4651.0	6829.9	152.32	23.93	41.50	323
8.94333	2345	0.1069	4799.8	7036.1	153.07	23.95	41.03	325
8.72223	2408	0.1033	4946.8	7239.8	153.79	23.97	40.56	328
8.51401	2471	0.0997	5092.3	7441.4	154.49	23.98	40.08	330
8.31737	2528	0.0968	5236.5	7641.5	155.17	24.00	39.82	332
8.13115	2589	0.0941	5379.2	7838.9	155.84	24.02	39.53	335

SE BOUNDARY

TABLE 16. THERMODYNAMIC PROPERTIES OF FLUORINE

21 MN/m² ISOBAR

TEMPERATURE (IPTS 1968) K	DENSITY MOL/L	ISOTHERM DERIVATIVE J/MOL	ISOCHORE DERIVATIVE MN/m²-K	INTERNAL ENERGY J/MOL	ENTHALPY J/MOL	ENTROPY J/MOL-K	Cv	Cp	VELOCITY OF SOUND M/S
							J / MOL - K		
* 55.493	45.2694	31861	4.3526	-6461.9	-5998.0	61.32	36.02	52.12	1102
56	45.1936	31366	4.3512	-6436.4	-5971.7	61.81	35.92	52.47	1098
58	44.8985	29458	4.3458	-6335.9	-5868.1	63.68	35.56	54.00	1085
60	44.5914	29065	4.3401	-6233.6	-5762.6	65.52	35.26	54.82	1090
62	44.2919	29022	4.2762	-6131.5	-5657.4	67.30	35.37	55.28	1093
64	44.0001	28279	4.1890	-6028.9	-5551.7	69.02	35.64	56.15	1083
66	43.6994	26815	3.9457	-5924.6	-5444.0	70.73	35.70	55.76	1050
68	43.3924	25522	3.7434	-5819.7	-5335.8	72.39	35.16	54.99	1025
70	43.0894	25223	3.6283	-5716.4	-5229.0	73.99	34.42	54.10	1021
72	42.7976	24620	3.5593	-5615.3	-5124.6	75.51	33.85	54.07	1017
74	42.4934	24187	3.5051	-5513.0	-5018.8	77.00	33.70	54.52	1015
76	42.1924	22557	3.3873	-5410.0	-4912.3	78.47	33.63	55.34	988
78	41.8844	21728	3.2531	-5306.0	-4804.7	79.91	33.41	55.06	971
80	41.5772	20943	3.0966	-5202.9	-4697.8	81.30	33.15	54.34	951
82	41.2668	20739	3.0012	-5100.1	-4591.2	82.66	32.82	53.74	945
84	40.9588	18975	2.9218	-4997.4	-4484.7	83.98	32.53	55.06	919
86	40.6429	18436	2.8054	-4893.8	-4377.1	85.29	32.16	54.38	906
88	40.3243	18081	2.7090	-4790.2	-4269.4	86.56	31.65	53.61	898
90	40.0146	16826	2.6321	-4688.8	-4164.0	87.78	31.48	54.62	877
92	39.6924	16551	2.5543	-4585.9	-4056.9	89.00	31.39	54.41	869
94	39.3711	16039	2.4829	-4483.0	-3949.6	90.18	31.13	54.43	859
96	39.0505	15097	2.4207	-4380.3	-3842.5	91.34	30.60	55.03	845
98	38.7242	14516	2.3621	-4277.4	-3735.1	92.48	30.23	55.35	836
100	38.3953	13892	2.3003	-4174.0	-3627.0	93.60	29.99	55.83	825
102	38.0629	13435	2.2297	-4070.0	-3518.3	94.70	29.85	55.91	814
104	37.7281	12973	2.1387	-3965.5	-3408.9	95.79	29.74	55.50	798
106	37.3971	12062	2.0601	-3861.9	-3300.3	96.84	29.29	55.92	779
108	37.0527	11619	2.0153	-3756.8	-3190.0	97.89	28.90	56.40	773
110	36.7087	11121	1.9442	-3651.6	-3079.6	98.92	28.79	56.54	758
112	36.3607	10611	1.8763	-3545.9	-2968.4	99.94	28.61	56.72	744
114	36.0070	10428	1.8207	-3439.3	-2856.1	100.94	28.52	56.48	737
116	35.6570	9781	1.7475	-3332.7	-2743.7	101.92	28.44	56.92	718
118	35.2977	9370	1.6803	-3224.9	-2630.0	102.90	28.29	56.83	704
120	34.9354	8883	1.6201	-3116.7	-2515.6	103.86	28.10	57.15	690
122	34.5679	8390	1.5639	-3007.6	-2400.1	104.81	27.98	57.74	675
124	34.1941	8055	1.5067	-2897.4	-2283.3	105.75	27.95	57.83	662
126	33.8178	7647	1.4502	-2786.6	-2165.6	106.67	28.07	58.37	647
128	33.4366	7278	1.3946	-2673.8	-2045.8	107.59	27.95	58.55	633
130	33.0506	6890	1.3400	-2560.0	-1924.6	108.50	27.83	58.84	619
132	32.6590	6522	1.2890	-2444.2	-1801.2	109.41	27.45	58.98	607
134	32.2619	6166	1.2377	-2327.4	-1676.5	110.29	27.27	59.26	594
136	31.8591	5815	1.1884	-1689.5	-1030.3	111.17	27.00	59.55	581
138	31.4473	5544	1.1434	-1578.9	-911.2	112.04	26.59	59.50	571
140	31.0318	5240	1.0970	-1468.7	-792.0	112.90	26.33	59.72	559
142	30.6117	4922	1.0530	-1358.5	-672.5	113.75	26.22	60.36	546
144	30.1812	4666	1.0205	-1246.9	-551.1	114.60	26.30	61.58	536
146	29.7499	4408	0.9779	-1134.3	-428.4	115.44	26.60	62.39	522
148	29.3048	4167	0.9227	-1020.1	-303.5	116.29	26.86	62.07	503
150	28.8615	3891	0.8755	-907.4	-179.8	117.12	26.76	62.24	488
152	28.4062	3673	0.8448	-793.7	-54.4	117.95	26.70	63.30	479
154	27.9436	3466	0.8059	-678.8	72.7	118.78	26.72	63.68	466
156	27.4733	3273	0.7658	-563.8	200.6	119.61	26.52	63.96	454
158	26.9964	3092	0.7357	-448.9	329.0	120.43	26.22	64.17	446
160	26.5146	2922	0.7107	-333.5	458.6	121.24	26.13	65.47	439
165	25.2914	2534	0.6281	-48.1	782.2	123.23	24.96	65.12	417
170	24.0419	2225	0.5587	238.0	1111.4	125.20	25.25	66.51	393
175	22.7867	1982	0.4962	526.1	1447.7	127.15	25.62	67.49	371
180	21.5434	1805	0.4423	809.4	1784.2	129.05	24.93	66.95	357
185	20.3414	1676	0.3942	1084.3	2116.7	130.87	24.19	65.64	346
190	19.1955	1597	0.3538	1351.3	2445.3	132.62	23.94	64.37	336
195	18.1268	1553	0.3186	1605.8	2764.3	134.28	23.54	62.31	329
200	17.1225	1565	0.2944	1853.1	3079.6	135.87	23.62	61.38	327
205	16.2366	1554	0.2636	2088.3	3381.7	137.37	24.04	58.81	316
210	15.4267	1568	0.2421	2311.4	3672.7	138.77	24.10	57.08	313
215	14.6927	1599	0.2234	2524.6	3953.9	140.09	24.13	55.21	310
220	14.0279	1642	0.2077	2728.5	4225.5	141.34	24.10	53.46	310
225	13.4266	1694	0.1938	2924.1	4488.1	142.52	24.05	51.74	310
230	12.8812	1750	0.1819	3112.2	4742.5	143.64	24.00	50.21	310
235	12.3854	1812	0.1713	3294.0	4989.5	144.70	23.95	48.77	312
240	11.9334	1873	0.1620	3469.9	5229.7	145.72	23.91	47.52	313
245	11.5130	1960	0.1559	3642.2	5466.2	146.69	23.90	46.83	318
250	11.1334	2026	0.1471	3809.6	5695.9	147.62	23.92	45.46	318
255	10.7857	2077	0.1394	3972.8	5919.8	148.51	23.93	44.43	319
260	10.4617	2132	0.1334	4132.6	6140.0	149.36	23.93	43.76	320
265	10.1572	2204	0.1289	4290.0	6357.5	150.19	23.93	43.31	324
270	9.87520	2268	0.1234	4444.8	6571.3	150.99	23.95	42.54	326
275	9.61271	2327	0.1183	4597.2	6781.8	151.76	23.97	41.87	327
280	9.36628	2386	0.1144	4747.6	6989.7	152.51	23.99	41.50	330
285	9.13407	2449	0.1102	4896.2	7195.3	153.24	24.01	40.95	332
290	8.91566	2511	0.1052	5043.0	7398.4	153.95	24.02	40.12	332
295	8.71066	2558	0.1025	5188.3	7599.1	154.63	24.04	40.00	335
300	8.51548	2615	0.0997	5332.2	7798.3	155.30	24.06	39.79	337

* TWO-PHASE BOUNDARY

XIII. ACKNOWLEDGEMENTS

The support for this work was furnished by the Air Force

ket Propulsion Laboratory, Edwards, California.

The authors wish to thank Professor K. D. Timmerhaus

his helpful comments and discussions concerning this task.

nowledgement is made to Dr. L. A. Weber for his many valuable

estions concerning the correlation of the data. Helpful com-

ts from Mr. D. E. Diller, Dr. R. D. Goodwin, and Dr. B. A.

nglove are appreciated very much. The assistance by Mr. R. D

arty and Mr. J. G. Hust in debugging some of the computer

rams is acknowledged.

Finally, thanks are due to Mr. G. K. Johnson of

nne National Laboratory for supplying the purified fluorine

for these experiments.

XIV. BIBLIOGRAPHY

(1)   Sengers, J. M. H. Levelt, "The Experimental Determina-
       tion of the Equation-of-State of Gases and Liquids at Low
       Temperatures," Physics of High Pressures and the Con-
       densed Phase, North-Holland Publishing Company,
       Amsterdam (1965), pp. 60-97.

(2)   Burnett, E. S., "Compressibility Determinations Without
       Volume Measurements," J. Appl. Mechanics, Trans. ASME,
       58, A136 (1936).

(3)   Goodwin, R. D., "Apparatus for Determination of Pressure-
       Density-Temperature Relations and Specific Heats of Hydro-
       gen to 350 Atmospheres at Temperatures Above 14°K," J.
       Research Nat. Bur. Standards 65C, No. 4, 231-43 (1961).

(4)   Weber, L. A., "P-V-T, Thermodynamic and Related Pro-
       perties of Oxygen from the Triple Point to 300 K at Pres-
       sures to 33 MN/m², " J. Research Nat. Bur. Standards,
       74A, No. 1, 93-129 (1970).

(5)   Straty, G. C. and Prydz, R., "Fluorine Compatible Appara-
       tus for Accurate PVT Measurements," Rev. Sci. Inst.,
       accepted for publication.

(6)   Straty, G. C., "Bellows-Sealed Valve for Reactive Gases at
       Moderately High Pressures," Rev. Sci. Inst. 40, No. 2,
       378-9 (1969).

(7)   Lashof, T. W. and Macurdy, L. B., "Precision Laboratory
       Standards of Mass and Laboratory Weights," Nat. Bur.
       Standards Circular 547 (1954).

(8)   Cross, J. L., "Reduction of Data for Piston Gage Pressure
       Measurements," Nat. Bur. Standards Monograph 65 (1963).

(9)   Smith, B. K., "Low Thermal Soldering Procedures for
       Copper Junctions," Electronic Design 7 (25), 52 (1959).

(10)   Goodwin, R. D., "An Improved D.C. Power Regulator,"
       Advances in Cryogenic Engineering, Vol. 5, 577-9, Plenum
       Press, New York (1960).

(11) Hoge, H. J., "Vapor Pressure and Fixed Points of Oxygen and Heat Capacity in the Critical Region," J. Research Nat. Bur. Standards 44, 321-45 (1950); RP 2081.

(12) Abramowitz, M., and Slegun, I. A., Handbook of Mathematical Functions with Formulas, Graphs, and Mathematical Tables, Nat. Bur. Standards Applied Mathematics Series No. 55 (June 1964) p. 879.

(13) Bedford, R. E., Durieux, M., Muijlwijk, R., and Barber, C. R., "Relationship between the International Practical Temperature Scale of 1968 and Various "National Temperature Scales" in the Range 13.81 K to 90.188 K," Metrologia 5, 47 (1969).

(14) -----------, "International Practical Temperature Scale of 1968, Adopted by the Comité International des Poids et Measures," Metrologia 5, 35 (1969).

(15) Powell, R. L., and Blanpied, W. A., "Thermal Conductivity of Metals and Alloys at Low Temperatures," Nat. Bur. Standards Circular 556 (1954).

(16) Prydz, R., Timmerhaus, K. D., and Stewart, R. B., "The Thermodynamic Properties of Deuterium," Advances in Cryogenic Engineering, Vol. 13, 384-96, Plenum Press, New York (1967).

(17) Rubin, T., Altman, H. W., and Johnston, H. L., "Coefficients of Thermal Expansion of Solids at Low Temperatures, J. Am. Chem. Soc. 76, 5289 (1954).

(18) Wassenaar, T., Ph.D. Thesis, Amsterdam Univ., Amsterdam (1952).

(19) Forsythe, W. E., Smithsonian Physical Tables, Ninth revised ed. (Smithsonian Institution, Washington, D.C., 1956).

(20) Roark, R. J., Formulas for Stress and Strain, 4th ed., McGraw-Hill, New York (1965), p. 308.

(21) Hoge, H. J., and Arnold, R. D., "Vapor Pressures of Hydrogen, Deuterium and Hydrogen Deuteride and Dew Point Pressures of Their Mixtures," J. Research Nat. Bur. of Standards 47, 63-74 (1951);RP 2228.

(22) Prydz, R., Straty, G. C., and Timmerhaus, K. D.,
      "Properties of Fluorine along the Vapor-Liquid Coexistence
      Boundary," to be published.

(23) Hust, J. G., and McCarty, R. D., "Curve-Fitting Techni-
      ques and Applications to Thermodynamics," Cryogenics, 7,
      No. 4, 200-06 (Aug. 1967).

(24) Goodwin, R. D., "Estimation of Critical Constants T_c, ρ_c
      from the $\rho(T)$ and $T(\rho)$ Relations at Coexistence," J. Res.
      Nat. Bur. Std., Vol. 74A, No. 2, 221-7 (1970).

(25) Prydz, R., this laboratory, unpublished.

(26) Kanda, E., "Studies on Fluorine at Low Temperatures. VII.
      Determination of Dielectric Constants of Condensed Gases,"
      Bull. Chem. Soc. Japan, 12, 473-9 (1937).

(27) Kilner, S. B., Randolph, C. L., and Gillespie, R. W.,
      "The Density of Liquid Fluorine," J. Am. Chem. Soc., 74,
      1086-7 (1952).

(28) Dunn, L. G., and Millikan, C. B., "Combined Bimonthly
      Summary, No. 29," Cal. Inst. Tech., Jet Propulsion Lab.,
      4 (May 20, 1952).

(29) Elverum, G. W. Jr., and Doescher, R. N., "Physical
      Properties of Liquid Fluorine," J. Chem. Phys., 20,
      1834-6 (Dec. 1952).

(30) White, D., Hu, J-H., and Johnston, H. L., "The Density
      and Surface Tension of Liquid Fluorine between 66 and 80 K,"
      J. Am. Chem. Soc., 76, 2584-6 (1954).

(31) Jarry, R. L., and Miller, H. C., "The Density of Liquid
      Fluorine between 67 and 103 K," J. Am. Chem. Soc., 78,
      1552-3 (1956).

(32) Moissan, H., and Dewar, J., "Nouvelles Expériences sur
      la Liquéfaction du Fluor," Comptes Rendus, 125, 505-11
      (1897).

(33) Drugman, J., and Ramsay, W., "Specific Gravities of the
      Halogens at their Boiling Points, and of Oxygen and Nitro-
      gen," J. Chemical Soc. (London), 77, 1228-33 (1900).

'hermodynamics of Oxygen, Hydrazine, and Fluorine,"
No. F-5028-101, Republic Aviation Corp., (1957).

· and Adamson, Tech. Bull. TA-85411, General
ical Division, Allied Chemical and Dye Corp., (1961).

G. H., and Hildebrand, J. H., "The Vapor Pressure
ritical Temperature of Fluorine," J. Am. Chem. Soc.,
339-43 (1930).

r, G. C., and Prydz, R., "Melting Curve and Triple-
Properties of Fluorine," Physics Letters A, 31A,
, 301-2 (1970).

-H., White, D., and Johnston, H. L., "Condensed
:alorimetry. V. Heat Capacities, Latent Heats and
pies of Fluorine from 13 to 85 K; Heats of Transition,
n, Vaporization and Vapor Pressures of the Liquid,"
ı. Chem. Soc., 75, 5642-5 (1953).

ı, E., "Studies on Fluorine at Low Temperatures. VIII.
mination of Molecular Heat, Heat of Fusion of Con-
d Fluorine and Entropy of Fluorine," Bull. Chem.
Japan, 12, 511-20 (1937).

na, S., and Kanda, E., "Studies on Fluorine at Low
eratures. II. Vapour Pressure of Fluorine," Bull.
. Soc. Japan, 12, 416-9 (1937).

sen, W. H., "The Vapor Pressure of Fluorine," J.
Chem. Soc., 56, 614-15 (1934).

·, G. C., and Prydz, R., "The Vapor Pressure of
l Fluorine," Advances in Cryogenic Engineering, Vol.
lenum Press, New York, 36-41 (1970).

ər, W. T., and Mullins, J. C., "Calculation of the
· Pressure and Heats of Vaporization and Sublimation
_uids and Solids, Especially Below One Atmosphere.
itrogen and Fluorine." Contract No. CST-7404,
ıal Bureau of Standards, Boulder, Colorado (April
)63).

(45) Goodwin, R. D., "Nonanalytic Vapor Pressure Equation With Data for Nitrogen and Oxygen," J. Res. Nat. Bur. Std. 73A, No. 5, 487-91 (1969).

(46) -----------, Janaf Interim Thermochemical Tables, The Dow Chemical Company, Midland, Michigan, December 31, 1960.

(47) Landau, R., and Rosen, R., Industrial Handling of Fluorine. Preparation, Properties and Technology of Fluorine and Organic Fluoro Compounds, C. Slesser and S. R. Schram, eds., McGraw-Hill Book Co., 1951, 133-57.

(48) Prydz, R., and Goodwin, R. D., "Specific Heats, C_v, of Compressed Liquid and Gaseous Fluorine," J. Res. Nat. Bur. Std., to be published.

(49) White, D., Hu, J-H., and Johnston, H. L., "The Inter-molecular Force Constants of Fluorine," J. Chem. Phys., 21, No. 7, 1149-52 (1953).

(50) Gosman, A. L., McCarty, R. D., and Hust, J. G., "Thermodynamic Properties of Argon from the Triple Point to 300 K at Pressures to 1000 Atmospheres," Nat. Std. Ref. Data Series - Nat. Bur. Std. 27, Washington, D. C. (Mar. 1969).

(51) Dibeler, V. H., Walker, J. A., and McCullah, K. E., "Photoionization Study of the Dissociation Energy of Fluorine and the Heat of Formation of Hydrogen Fluoride," J. Chem. Phys., 51, 4230-5 (Nov. 1969).

(52) Mayer, J. E., and Mayer, M. G., Statistical Mechanics, John Wiley & Sons, Inc., New York (1940).

(53) Herzberg, G., Molecular Spectra and Molecular Structure, D. Van Nostrand Company, Inc., New York (1950).

(54) Andrychuk, D., "The Raman Spectrum of Fluorine," Can. J. Phys., 29, 151-8 (1951).

(55) Hust, J. G., and Gosman, A. L., "Functions for the Cal-culations of Entropy, Enthalpy, and Internal Energy for Real Fluids Using Equations of State and Specific Heats," Advances in Cryogenic Engineering, Vol. 9, 227-33, Plenum Press, New York (1964).

Prydz, R., "The Thermodynamic Properties of Deuterium," M.S. Thesis, University of Colorado, Boulder, Colorado (1967).

Goodwin, R. D., and Prydz, R., "Specific Heats of Fluorine at Coexistence," J. Res. Nat. Bur. Std., to be published.

Yang, C. N., and Yang, C. P., "Critical Point in Liquid Gas Transitions," Phys. Rev. Letters, 13 (9), 303-5 (1964).

Barmatz, M., "Ultrasonic Measurements Near the Critical Point of He^4," Phys. Rev. Letters, 24 (12), 651-4 (1970).

Grigor, A. F., and Steele, W. A., "Density Balance for Low Temperatures and Elevated Pressures," Rev. Sci. Inst. 37, No. 1, 51-4 (1966).

Goldman, K., and Scrase, N. G., "Densities of Saturated Liquid Oxygen and Nitrogen," Physica 44, No. 4, 555-86 (1969).

Haynes, W. M., "A Magnetic Densitometer for Cryogenic Fluids," University of Virginia, Charlottesville, Virginia, Ph.D. Dissertation (June 1970).

Weber, L. A., "The P-V-T Surface of Oxygen in the Critical Region; Densities of Saturated Liquid and Vapor," Advances in Cryogenic Engineering, Vol. 15, 50-7, Plenum Press, New York (1970).

Straty, G. C., this laboratory, personal communications (March 1970).

APPENDIX A - TEMPERATURE SCALE DIFFERENCES

Given below are temperature scale differences from References [13] (T, IPTS 1968 - T, NBS 1955) and [14] (T, IPTS 1968 - T, IPTS 1948) for temperatures below and above 90 K, respectively.

T, IPTS 1968 - K	T, IPTS 1968 - T, old - K	T, IPTS 1968 - K	T, IPTS 1968 - T, old - K
50.0	0.0131	83.0	-0.0064
51.0	0.0124	84.0	-0.0049
52.0	0.0116	85.0	-0.0029
53.0	0.0104	86.0	-0.0005
54.0	0.0089	87.0	0.0022
55.0	0.0071	88.0	0.0049
56.0	0.0052	89.0	0.0074
57.0	0.0034	90.0	0.0096
58.0	0.0017		
59.0	0.0003	93.15	0.012
60.0	-0.0008	103.15	0.007
61.0	-0.0014	113.15	-0.005
62.0	-0.0015	123.15	-0,013
63.0	-0.0012	133.15	-0.013
64.0	-0.0007	143.15	-0.006
65.0	-0.0001	153.15	0.003
66.0	0.0005	163.15	0.013
67.0	0.0009	173.15	0.022
68.0	0.0011	183.15	0.029
69.0	0.0009	193.15	0.033
70.0	0.0003	203.15	0.034
71.0	-0.0006	213.15	0.032
72.0	-0.0017	223.15	0.029
73.0	-0.0030	233.15	0.024
74.0	-0.0043	243.15	0.018
75.0	-0.0056	253.15	0.012
76.0	-0.0068	263.15	0.006
77.0	-0.0078	273.15	0.000
78.0	-0.0086	283.15	-0.004
79.0	-0.0090	293.15	-0.007
80.0	-0.0090	303.15	-0.009
81.0	-0.0086	313.15	-0.010
82.0	-0.0077	323.15	-0.010

CPSIA information can be obtained
at www.ICGtesting.com
Printed in the USA
BVHW040958071218
534846BV00028B/278/P